Studies in Universal Logic

This series is devoted to the universal approach to logic and the development of a general theory of logics. It covers topics such as global set-ups for fundamental theorems of logic and frameworks for the study of logics, in particular logical matrices, Kripke structures, combination of logics, categorical logic, abstract proof theory, consequence operators, and algebraic logic. It includes also books with historical and philosophical discussions about the nature and scope of logic. Three types of books will appear in the series: graduate textbooks, research monographs, and volumes with contributed papers.

Witold A. Pogorzelski
Piotr Wojtylak

Completeness Theory for Propositional Logics

Birkhäuser
Basel · Boston · Berlin

Authors:

Witold A.Pogorzelski
Institute of Mathematics
University of Bialystok
Akademicka 2
15-275 Bialystok
Poland

Piotr Wojtylak
Institute of Mathematics
Silesian University
Bankowa 14
40-007 Katowice
Poland
e-mail: wojtylak@us.edu.pl

2000 Mathematical Subject Classification: 03-01, 03B05, 03G10

Library of Congress Control Number: 2008923069

Bibliographic information published by Die Deutsche Bibliothek
Die Deutsche Bibliothek lists this publication in the Deutsche Nationalbibliografie;
detailed bibliographic data is available in the Internet at <http://dnb.ddb.de>.

ISBN 978-3-7643-8517-0 Birkhäuser Verlag AG, Basel · Boston · Berlin

© 2008 Birkhäuser Verlag AG
Basel · Boston · Berlin
P.O. Box 133, CH-4010 Basel, Switzerland
Part of Springer Science+Business Media
Printed on acid-free paper produced from chlorine-free pulp. TCF ∞

ISBN 978-3-7643-8517-0

ISBN 978-3-7643-8518-7 (eBook)

9 8 7 6 5 4 3 2 1

www.birkhauser.ch

Contents

Introduction

Completeness is one of the most important notions in logic and the foundations of mathematics. Many variants of the notion have been defined in literature. We shall concentrate on these variants, and aspects, of completeness which are defined in propositional logic.

Completeness means the possibility of getting all correct and reliable schemata of inference by use of logical methods. The word 'all', seemingly neutral, is here a crucial point of distinction. Assuming the definition as given by E. Post we get, say, a global notion of completeness in which the reliability refers only to syntactic means of logic and outside the correct schemata of inference there are only inconsistent ones. It is impossible, however, to leave aside local aspects of the notion when we want to make it relative to some given or invented notion of truth. Completeness understood in this sense is the adequacy of logic in relation to some semantics, and the change of the logic is accompanied by the change of its semantics. Such completeness was effectively used by J. Łukasiewicz and investigated in general terms by A. Tarski and A. Lindenbaum, which gave strong foundations for research in logic and, in particular, for the notion of consequence operation determined by a logical system.

The choice of logical means, by use of which we intend to represent logical inferences, is also important. Most of the definitions and results in completeness theory were originally developed in terms of propositional logic. Propositional formal systems find many applications in logic and theoretical computer science. Due to the simplicity of the language, one can use in research various methods and results of abstract algebra and lattice theory. Propositional completeness theory is a prerequisite for the other types of completeness theory and its applications.

In this monograph we wish to present a possibly uniform theory of the notion of completeness in its principal version, and to propose its unification and, at the same time, generalization. This is carried out through the definition and analysis of the so-called Γ-completeness (Γ is any set of propositional formulas) which generalizes and systematizes some variety of the notion of completeness for propositional logics — such as Post-completeness, structural completeness and many others. Our approach allows for a more profound view upon some essential properties (e.g., two-valuedness) of propositional systems. For these purposes we shall use, as well, the elementary means of general algebra, the theory of logical matrices, and the theory of consequence operations.

The subject of completeness became a separate area of research in propositional logic in the 1970s. Results of research of many authors, mainly Polish, have been used here. Our exposition is based on a former manuscript [88], 1982. We have tried to include all important results which are indispensable for further work in the area. In addition, we have included some of the more recent results stimulating present research. The book is organized on the following plan. Basic methods and constructions of universal algebra and propositional logic are briefly discussed in Chapter 1. Main results are exposed in the next two chapters; Chapter 2 deals with local, and Chapter 3 with global, aspects of the notion of completeness. In the last chapter and appendices we present some more advanced topics which combine several methods and ideas involved in previous fragments. The terminology and notation employed in our monograph is standard. The set theoretical symbols $\emptyset$, $\in$, $\subseteq$, $\cap$, $\cup$ etc. have their usual meanings. The power set of the set A is denoted by 2^A and $Nc(A)$ is the cardinality of A. We use $\Rightarrow$, $\vee$, $\wedge$, $\neg$, $\Leftrightarrow$, $\forall$, $\exists$ for (meta)logical connectives and quantifiers whereas $\rightarrow$, $+$, $\cdot$, $\sim$, $\equiv$ are reserved for (intra)logical symbols, i.e., symbols of the considered logical systems.

Witold A. Pogorzelski
Piotr Wojtylak

Chapter 1

Basic notions

This chapter gives a concise background for the further study of propositional systems. We assume that the reader is familiar with elements of propositional logic and therefore some basic facts will be stated without proofs. Simple results will be often given without references.

1.1 Propositional languages

Any *propositional language* is determined by an infinite set *At* of *propositional variables* (it is usually assumed that *At* is denumerable, e.g., $At = \{p_0, p_1, \ldots\}$) and by a finite set of *propositional connectives* $F_1, F_2, \ldots, F_n$. Suppose that F_i, for $i \leqslant n$, is a k_i-ary connective (usually, we have $k_i \leqslant 2$). The set S of the *propositional formulas* is defined as follows:

(i) $At \subseteq S$,

(ii) $\alpha_1, \ldots, \alpha_{k_i} \in S \quad \Rightarrow \quad F_i(\alpha_1, \ldots, \alpha_{k_i}) \in S$, for each $i \leqslant n$.

Formulas will be denoted by Greek letters $\alpha, \beta, \ldots$. The number of occurrences of the signs $F_1, \ldots, F_n$ in a formula $\alpha \in S$ will be called *the length of* α; the precise definition of the length is as follows:

(i) $l(\gamma) = 0 \quad \text{if} \quad \gamma \in At$,

(ii) $l(F_i(\alpha_1, \ldots, \alpha_k)) = l(\alpha_1) + \ldots + l(\alpha_k) + 1$.

Similarly, we define, for each $\alpha \in S$, the set $Sf(\alpha)$ of *subformulas* and the set $At(\alpha)$ of *variables occurring* in α:

(i) $Sf(\gamma) = At(\gamma) = \{\gamma\} \quad \text{if} \quad \gamma \in At$,

(ii) $At(F_i(\alpha_1, \ldots, \alpha_k)) = At(\alpha_1) \cup \ldots \cup At(\alpha_k)$,

(iii) $Sf(F_i(\alpha_1, \ldots, \alpha_k)) = Sf(\alpha_1) \cup \ldots \cup Sf(\alpha_k) \cup \{F_i(\alpha_1, \ldots, \alpha_k)\}$.

If we write $\alpha(p_0, \ldots, p_k)$, it will mean that $At(\alpha) \subseteq \{p_0, \ldots, p_k\}$. Moreover, let $At(X)$, for any $X \subseteq S$, be the set of the variables occurring in X, i.e.,

$$At(X) = \bigcup \{At(\alpha) : \alpha \in X\}.$$

The family of all finite subsets of the set X will be denoted by $\mathrm{Fin}(X)$:

$$Y \in \mathrm{Fin}(X) \quad \equiv \quad Y \subseteq X \ \wedge \ Y \text{ is finite};$$

$$Y \in \mathrm{Fin}^*(X) \quad \equiv \quad Y \neq \emptyset \ \wedge \ Y \in \mathrm{Fin}(X).$$

Formal languages together with inferential rules originated in the efforts of logicians to state precisely the intuitive notion of a proof. *Inferential rules* were introduced in order to represent elementary links of deductive reasoning which constitute (when used step by step) any mathematical proof. For a long time mathematical logic remained devoted to this original problem, and it was appropriate that its methods were restricted to those possessing a finitistic character. It was not till several years later that there appeared new problems and new points of view which led to the introduction of some infinitistic methods.

In the case of inferential rules, it has been marked by the consideration of rules with infinite sets of premises. It should be stressed that we are not interested in the development of infinitistic methods here. We focus on the finitistic point of view and therefore languages with infinitely many connectives as well as infinitary formulas will be left outside the scope of the present book.

The development of mathematical logic and the application of elements of universal algebra forced us, however, to consider non-denumerable languages and infinitary rules to some extent. Nevertheless, we do not want to miss advantages which arise when only finitary rules are considered. Therefore, our further examinations will concern the finitistic as well as the infinitistic point of view. We will examine two different definitions of the notion of a rule; however, the set of all rules over S will be denoted, in both cases, as $\mathscr{R}_S$ (or $\mathscr{R}$ if this is not misleading).

Definition 1.1.

(fin) $r \in \mathscr{R}_S \quad \Leftrightarrow \quad r \subseteq \mathrm{Fin}^*(S) \times S$;

(∞) $r \in \mathscr{R}_S \quad \Leftrightarrow \quad r \subseteq 2^S \times S$.

The results to come which refer to Definition 1.1 (fin) will be indicated by the sign (fin), those referring to 1.1 (∞) by (∞). The lack of the indices (i.e., (∞) and (fin)) will mean that the statement is true for finitary as well as for arbitrary (possible infinitary) rules.

Thus, rules of inference are understood as families of *sequents* $\langle \Pi, \alpha \rangle$, where Π is a set of formulas ($\Pi \subseteq S$) and $\alpha \in S$. If $\langle \Pi, \alpha \rangle \in r$, then α is said to be a *conclusion* obtained from the *premises* Π by use of the rule r. The notation of sequents $\langle \Pi, \alpha \rangle \in r$ in the form of inferential schemata

$$r : \frac{\Pi}{\alpha}$$

is very suggestive and will often be used.

Note that, according to 1.1 (fin), sequents $\langle \emptyset, \alpha \rangle$ are out of the reach of any rule. However, such rules are not excluded by 1.1 (∞). *Axiomatic rules*, i.e., rules with empty sets of premises may be regarded as sets of formulas.

In our further considerations we will mainly use logical systems formalized in the *standard propositional language* provided by *the truth-functional connectives* $\rightarrow$ (implication), $+$ (disjunction), $\cdot$ (conjunction) and $\sim$ (negation). We will also use the abbreviation $\alpha \equiv \beta$ to denote the formula $(\alpha \rightarrow \beta) \cdot (\beta \rightarrow \alpha)$. The set of all formulas built up from *At* by means of the connectives $\rightarrow$, $+$, $\cdot$, $\sim$ will be denoted as S_2. The symbol S_1, where $S_1 \subseteq S_2$, will stand for the set of all positive (without negation) expressions. We will consider, as well, other sublanguages of S_2.

Let us define, as an example of a rule, *the modus ponens rule*

$$r_0 = \{ \langle \{\alpha \rightarrow \beta, \alpha\}, \beta \rangle : \ \alpha, \beta \in S_2 \}.$$

Defining r_0 by use of a schema one can write

$$r_0 : \frac{\alpha \rightarrow \beta \ , \ \alpha}{\beta} \qquad \text{for all} \ \alpha, \beta \in S_2.$$

Brackets in formulas will be omitted where convenient, the convention for reading formulas being that $\sim$ binds more strongly than $+$ and $\cdot$, the latter binding more strongly than $\rightarrow$, $\equiv$.

1.2 Abstract algebras

This section contains elements of the theory of abstract algebras. The results are presented from the point of view of their applications to logic.

If A is a non-empty set, then any mapping

$$f : A^k \rightarrow A, \quad k \geqslant 0$$

is called a *k-ary operation* on A. A system $\mathscr{A} = \langle A, f_1, \ldots, f_n \rangle$ is called an *(abstract) algebra* if the set A (the universe of $\mathscr{A}$) is non-empty and $f_1, \ldots, f_n$ are operations on A. Algebras will be denoted by script letters $\mathscr{A}$, $\mathscr{B}$, $\mathscr{C}$, $\ldots$ and the universes by the corresponding capital letters A, B, C, $\ldots$.

Let $\mathscr{A} = \langle A, f_1, \ldots, f_n \rangle$ and $\mathscr{B} = \langle B, g_1, \ldots, g_n \rangle$ be two similar algebras, i.e., algebras of the same type. A mapping $h : A \rightarrow B$ such that

$$h\big(f_i(a_1, \ldots, a_k)\big) = g_i\big(h(a_1), \ldots, h(a_k)\big), \quad \text{for all} \ i \leqslant n \ \text{and} \ a_1, \ldots, a_k \in A$$

is called *a homomorphism* (in symbols $h : \mathscr{A} \rightarrow \mathscr{B}$).

One-to-one homomorphisms are called *embeddings*. If h is one-to-one and onto, it is then an *isomorphism*. We say that $\mathscr{A}$ is embeddable in $\mathscr{B}$ if there is an embedding from $\mathscr{A}$ into $\mathscr{B}$. The algebras $\mathscr{A}$ and $\mathscr{B}$ are said to be isomorphic ($\mathscr{A} \cong \mathscr{B}$) iff there exists an isomorphism h from $\mathscr{A}$ onto $\mathscr{B}$.

Let $\mathscr{A} = \langle A, f_1, \ldots, f_n \rangle$ and $\mathscr{B} = \langle B, g_1, \ldots, g_n \rangle$. Then $\mathscr{B}$ is called a *subalgebra* of $\mathscr{A}$ ($\mathscr{B} \subseteq \mathscr{A}$) iff $B \subseteq A$ and g_i (for each $i \leqslant n$) is the restriction of f_i to the set B.

Lemma 1.2. *If $h : A \to B$ is a homomorphism from $\mathscr{A}$ into $\mathscr{B}$, then $h(A)$ is closed under the operations of $\mathscr{B}$ (i.e., is the universe of a subalgebra of $\mathscr{B}$).*

Obviously, $\mathscr{A}$ is embeddable in $\mathscr{B}$ iff $\mathscr{A}$ is isomorphic with some subalgebra of $\mathscr{B}$. If $\mathscr{B}$ is a subalgebra of $\mathscr{A}$, then a non-empty set H is said to be *a generating set* of $\mathscr{B}$ iff $\mathscr{B}$ is the smallest subalgebra of $\mathscr{A}$ containing H. Any non-empty subset of the universe of $\mathscr{A}$ generates a subalgebra of $\mathscr{A}$. We say that H is a set of generators of $\mathscr{A}$ iff there is no proper subalgebra of $\mathscr{A}$ containing H.

Lemma 1.3. *Let h and h_1 be homomorphisms from $\mathscr{A}$ into $\mathscr{B}$ and let H be a set of generators of $\mathscr{A}$. Then*

$$ h|_H = h_1|_H \quad \Rightarrow \quad h = h_1. $$

In other words, if a mapping $v : H \to B$ can be extended to a homomorphism $h^v : \mathscr{A} \to \mathscr{B}$, then this extension is unique. Obviously, it is not true that every mapping (defined on a set) of generators can be extended to a homomorphism.

Given a class $\mathbf{K}$ of similar algebras and an algebra $\mathscr{A} \in \mathbf{K}$ generated by a set H, the algebra is said to be *free over* $\mathbf{K}$, with the *free generating set* H iff each mapping $v : H \to B$, for each $\mathscr{B} \in \mathbf{K}$, can be extended to a homomorphism $h^v : \mathscr{A} \to \mathscr{B}$. By Lemma 1.3, the homomorphism h^v is uniquely defined.

Any propositional language determines an algebra $\mathscr{S} = \langle S, F_1, \ldots, F_n \rangle$ in which F_i ($1 \leqslant i \leqslant n$) denotes the operator of forming F_i-propositions.

Lemma 1.4. *The algebra $\mathscr{S}$ of a propositional language is free over the class of all similar algebras and the propositional variables are the free generators of $\mathscr{S}$.*

Therefore, if $\mathscr{A} = \langle A, f_1, \ldots, f_n \rangle$ is an algebra similar to $\mathscr{S}$, then each mapping $v : At \to A$ can be uniquely extended to a homomorphism $h^v : S \to A$.

Let $\mathscr{A} = \langle A, f_1, \ldots, f_n \rangle$ be an algebra and ϱ a binary relation on A. Then ϱ is called *a congruence relation* on $\mathscr{A}$ if it is an equivalence relation (i.e., reflexive, symmetric and transitive) satisfying the property

$$ a_1 \varrho b_1 \wedge \ldots \wedge a_k \varrho b_k \quad \Rightarrow \quad f_i(a_1, \ldots, a_k) \varrho f_i(b_1, \ldots, b_k) $$

for each $i \leqslant n$ and each $a_1, \ldots, a_k, b_1, \ldots, b_k \in A$.

For any algebra $\mathscr{A}$ and any congruence relation ϱ on $\mathscr{A}$ we can construct the so-called *quotient algebra*

$$ \mathscr{A}/\varrho = \langle A/\varrho, g_1, \ldots, g_n \rangle $$

with $A/\varrho = \{[a]_\varrho : a \in A\}$ and the operations g_i defined by

$$g_i([a_1]_\varrho, \ldots, [a_k]_\varrho) = [f_i(a_1, \ldots, a_k)]_\varrho.$$

Thus, the algebras $\mathscr{A}$ and $\mathscr{A}/\varrho$ are similar and the natural (canonical) mapping h_ϱ from A onto A/ϱ,

$$h_\varrho(a) = [a]_\varrho \quad \text{for each} \quad a \in A,$$

is a homomorphism from $\mathscr{A}$ onto $\mathscr{A}/\varrho$.

Lemma 1.5. *If h is a homomorphism from $\mathscr{A}$ onto $\mathscr{B}$, then the binary relation ϱ on A induced by h, i.e.,*

$$a \varrho b \quad \Leftrightarrow \quad h(a) = h(b),$$

is a congruence relation on $\mathscr{A}$ and the algebras $\mathscr{A}/\varrho$ and $\mathscr{B}$ are isomorphic.

Let $\{\mathscr{A}_t\}_{t \in T}$ be a family of similar algebras $\langle A_t, f_1^t, \ldots, f_n^t \rangle$. The *product* (or, more specifically, the *direct product*) of this family is the algebra

$$\mathbf{P}_{t \in T} \mathscr{A}_t = \langle \mathbf{P}_{t \in T} A_t, f_1, \ldots, f_n \rangle$$

where $\mathbf{P}_{t \in T} A_t$ is the Cartesian product of the sets A_t and

$$f_i(\langle a_1^t \rangle_{t \in T}, \ldots, \langle a_k^t \rangle_{t \in T}) = \langle f_i^t(a_1^t, \ldots, a_k^t) \rangle_{t \in T}.$$

In the above definition elements of the product are regarded as T-sequences $\langle a^t \rangle_{t \in T}$ such that $a^t \in A_t$ for each $t \in T$ and the operations $f_1, \ldots, f_n$ are defined componentwise. The product of two algebras $\mathscr{A}$ and $\mathscr{B}$ will be denoted by $\mathscr{A} \times \mathscr{B}$. If $\mathscr{A}_t = \mathscr{A}$ for each $t \in T$, then we will write $\mathscr{A}^T$ for the *power* of the algebra $\mathscr{A}$. Let $\mathscr{A}^n$ stand for $\mathscr{A}^{\{1,\ldots,n\}}$. The *projection* from the product onto the t-axis will be denoted by π_t, i.e.,

$$\pi_t(\langle a^s \rangle_{s \in T}) = a^t, \quad \text{for} \quad t \in T$$

and, obviously, π_t is a homomorphism from $\mathbf{P}_{t \in T} \mathscr{A}_t$ onto $\mathscr{A}_t$.

Lemma 1.6. *Let h_t, for each $t \in T$, be a homomorphism from $\mathscr{A}$ into $\mathscr{A}_t$. We define a mapping h from A into the product $\mathbf{P}_{t \in T} A_t$ by*

$$h(a) = \langle h_t(a) \rangle_{t \in T}.$$

The mapping h is a homomorphism from $\mathscr{A}$ into the product $\mathbf{P}_{t \in T} \mathscr{A}_t$ and we have $\pi_t \circ h = h_t$ for each $t \in T$.

The mapping h described in Lemma 1.6 is called the product of the family $\{h_t\}_{t \in T}$. It can be proved that any homomorphism h from $\mathscr{A}$ into $\mathbf{P}_{t \in T} \mathscr{A}_t$ is the product of the mappings $\{\pi_t \circ h\}_{t \in T}$. We can also get by Lemma 1.6:

Corollary 1.7. *Let $\{\mathscr{A}_t\}_{t\in T}$ and $\{\mathscr{B}_t\}_{t\in T}$ be two families of similar algebras and let h_t, for each $t \in T$, be a homomorphism from $\mathscr{A}_t$ into $\mathscr{B}_t$. Then the mapping*

$$h(\langle a^t \rangle_{t\in T}) = \langle h_t(a^t) \rangle_{t\in T}$$

is a homomorphism from $\mathbf{P}_{t\in T}\mathscr{A}_t$ into $\mathbf{P}_{t\in T}\mathscr{B}_t$.

Corollary 1.8. *Let $\{\mathscr{A}_t\}_{t\in T}$ and $\{\mathscr{B}_t\}_{t\in T}$ be two indexed families of similar algebras and assume that $\mathscr{A}_t$ is embeddable in $\mathscr{B}_t$ for each $t \in T$. Then the product $\mathbf{P}_{t\in T}\mathscr{A}_t$ is embeddable into the product $\mathbf{P}_{t\in T}\mathscr{B}_t$.*

One-element algebras are said to be *degenerate*.

Lemma 1.9. *For any algebra $\mathscr{A}$ and any natural numbers n, m:*

(i) $\mathscr{A}^n \times \mathscr{A}^m \cong \mathscr{A}^{n+m}$;

(ii) $(\mathscr{A}^n)^m \cong \mathscr{A}^{n\cdot m}$;

(iii) $\mathscr{A}^n$ *is embeddable in* $\mathscr{A}^{n+m}$;

(iv) $\mathscr{A} \times \mathscr{B} \cong \mathscr{A}$ *if $\mathscr{B}$ is degenerate.*

Proof. The properties (i), (ii), (iv) are quite obvious. We shall prove (iii) only. Assume that $n = 1$. Then an embedding $h : \mathscr{A} \to \mathscr{A}^{m+1}$ can be defined by

$$h(x) = \langle x, \ldots, x \rangle, \quad \text{for each} \ \ x \in A.$$

It is clear that h is a one-to-one homomorphism from $\mathscr{A}$ into $\mathscr{A}^{m+1}$. Let $n > 1$. Then it follows from (i) that

$$\mathscr{A}^n \cong \mathscr{A}^{n-1} \times \mathscr{A} \quad \text{and} \quad \mathscr{A}^{n+m} \cong \mathscr{A}^{n-1} \times \mathscr{A}^{m+1}.$$

Hence if suffices to show that $\mathscr{A}^{n-1} \times \mathscr{A}$ is embeddable in $\mathscr{A}^{n-1} \times \mathscr{A}^{m+1}$. Notice that $\mathscr{A}^{n-1}$ is embeddable in $\mathscr{A}^{n-1}$. Moreover, it has been proved that $\mathscr{A}$ is embeddable in $\mathscr{A}^{m+1}$. Thus, by Corollary 1.8, $\mathscr{A}^{n-1} \times \mathscr{A}$ is embeddable in $\mathscr{A}^{n-1} \times \mathscr{A}^{m+1}$. $\square$

1.3   Preliminary lattice-theoretical notions

The basic notions of lattice theory will be briefly introduced in this section. We assume that the reader has a working knowledge of elements of lattice theory, hence definitions of some elementary notions and proofs of many basic facts will be omitted (they can be found, for instance, in Grätzer [28] , 1978).

Lattices

An algebra $\langle A, \cup, \cap \rangle$, where $\cup$, $\cap$ are two binary operations on A, is called a *lattice* iff there is an ordering $\leqslant$ on A such that $x \cap y$ is the greatest lower bound — infimum — and $x \cup y$ is the least upper bound — supremum — (with respect to $\leqslant$) of the set $\{x, y\}$, i.e.,

$$x \cap y = \inf\{x, y\} \quad \text{and} \quad x \cup y = \sup\{x, y\}$$

for every $x, y \in A$. A lattice $\langle A, \cup, \cap \rangle$ is said to be *complete* iff $\sup(X)$ and $\inf(X)$ exist for each $X \subseteq A$.

The greatest element in the lattice $\mathscr{A}$, if it exists, is denoted by 1_A (or 1 if this is not misleading). The least element, if it occurs, is symbolized by 0_A (or 0). Clearly, every finite lattice contains the greatest and the least element.

A lattice $\langle A, \cup, \cap \rangle$ is *distributive* iff the following holds for each $x, y, z \in A$:

$$x \cap (y \cup z) = (x \cap y) \cup (x \cap z),$$
$$x \cup (y \cap z) = (x \cup y) \cap (x \cup z).$$

Assume that $\langle A, \cup, \cap \rangle$ is a lattice and let $x, y \in A$. The greatest element of the set $\{z \in A : x \cap z \leqslant y\}$, if it exists, is called the *relative pseudo-complement of x to y* and is denoted by $x \dotrightarrow y$. Note that there are finite lattices in which $x \dotrightarrow y$ does not exist for some x, y. However, finite distributive lattices contain the relative pseudo-complement of x to y for each $x, y \in A$.

If the element $x \dotrightarrow y$ exists for each $x, y \in A$, then $\langle A, \dotrightarrow, \cup, \cap \rangle$ is called an *implicative* lattice. It is easy to prove that each implicative lattice is distributive and contains the unit element $1 = x \dotrightarrow x$ for each $x \in A$.

Observe that there are implicative lattices without the least element (zero). The existence of the zero-element in an implicative lattice $\langle A, \dotrightarrow, \cup, \cap \rangle$ allows us to define the operation of *pseudo-complementation*

$$-x = x \dotrightarrow 0, \quad \text{for all } x \in A.$$

It can be easily seen that $z \leqslant -x$ iff $x \cap z = 0$, hence $-x$ is the greatest element of the set $\{z \in A : x \cap z = 0\}$.

An algebra $\langle A, \dotrightarrow, \cup, \cap, - \rangle$ with three binary operations $\dotrightarrow$, $\cup$, $\cap$ and one monadic operation $-$ is called a *Heyting (pseudo-Boolean) algebra* iff $\langle A, \dotrightarrow, \cup, \cap \rangle$ is an implicative lattice with the least element 0 and $-x = x \dotrightarrow 0$ for every $x \in A$.

Note that the distributivity laws hold in any Heyting algebra. Needless to say, every finite distributive lattice $\langle A, \cup, \cap \rangle$ contains the least element and, for each $x, y \in A$, the relative pseudo-complement $x \dotrightarrow y$ exists in A. Thus, any finite distributive lattice can be considered as a Heyting algebra.

Lemma 1.10. *In each Heyting algebra:*

(i) $x \leqslant y \Rightarrow -y \leqslant -x$;   (iv) $---x = -x$;

(ii) $x \cap -x = 0$;   (v) $-(x \cup y) = -x \cap -y$;

(iii) $x \leqslant --x$;   (vi) $-x \cup -y \leqslant -(x \cap y)$.

Let $\mathscr{A} = \langle A, \rightarrow, \cup, \cap, - \rangle$ be a non-degenerate (i.e., $1_A \neq 0_A$) Heyting algebra. We shall specify some elements of $\mathscr{A}$. An element $x \in A$ is said to be *dense* iff $-x = 0$. The set of all dense elements in the algebra $\mathscr{A}$ is denoted by $G(\mathscr{A})$. This set is not empty as $1 \in G(\mathscr{A})$. The dense elements possess, among others, the following properties:

Lemma 1.11. *In each Heyting algebra:*

(i) $x \in G(\mathscr{A})$ $\Leftrightarrow$ $--x = 1$;

(ii) $x \in G(\mathscr{A})$ $\Leftrightarrow$ $(x \cap y \neq 0, \ for \ each \ y \neq 0)$;

(iii) $x \in G(\mathscr{A})$ $\Leftrightarrow$ $(x = y \cup -y, \ for \ some \ y \in A)$.

This lemma results directly from the definition and from Lemma 1.10.

Any Heyting algebra in which $x \cup -x = 1$ for each $x \in A$ is called a *Boolean algebra*. The standard example of a Boolean algebra is $\langle 2^X, \rightarrow, \cup, \cap, - \rangle$ where $\cup$, $\cap$, $-$ coincide with the set theoretical operations and the lattice ordering coincides with the relation of inclusion. A classical example of a Boolean algebra is, as well, the two-element algebra $\mathscr{B}_2$, i.e., the lattice which contains exactly two elements 1 and 0 (and $0 < 1$). Note that all two-element Boolean algebras are isomorphic.

Let $\mathscr{A}$, $\mathscr{B}$ be two Heyting (Boolean) algebras and let h be a homomorphism from $\mathscr{A}$ into $\mathscr{B}$, i.e., $h : A \rightarrow B$ preserves the operations of the algebras. It is evident that h preserves, as well, the ordering relation, the unit- and the zero-element, i.e.,

$$x \leqslant_A y \quad \Rightarrow \quad h(x) \leqslant_B h(y); \qquad h(1_A) = 1_B; \qquad h(0_A) = 0_B.$$

Lemma 1.12. *Let $\mathscr{A}$ and $\mathscr{B}$ be Heyting (Boolean) algebras. Then a mapping h from A onto B is an isomorphism of the algebras iff*

$$x \leqslant_A y \quad \Leftrightarrow \quad h(x) \leqslant_B h(y), \qquad for \ all \ x, y \in A.$$

Sublattices

According to the general notion of a subalgebra it is evident that each subalgebra of a lattice (implicative lattice, Heyting or Boolean algebra) $\mathscr{A}$ is also a lattice (implicative lattice, etc.). If $\mathscr{B}$ is a subalgebra of a lattice $\mathscr{A}$ $(\mathscr{B} \subseteq \mathscr{A})$, then the ordering relation on $\mathscr{B}$ is induced from $\mathscr{A}$, i.e.,

$$x \leqslant_B y \quad \Leftrightarrow \quad x \leqslant_A y, \qquad for \ all \ x, y \in B.$$

If — in addition — $\mathscr{A}$ is a Heyting (Boolean) algebra, then $1_A = 1_B$ and $0_A = 0_B$.

Lemma 1.13. *Each finitely generated lattice contains the unit- and zero-element.*

Proof. Suppose that $\{x_1, \ldots, x_k\}$, where $k > 0$, is a set of generators of a lattice $\langle A, \cup, \cap \rangle$. Let us define a subset of the set A as follows:

$$K = \{x \in A : x_1 \cap \ldots \cap x_k \leqslant x \leqslant x_1 \cup \ldots \cup x_k\}.$$

Obviously, $x_i \in K$ for each $i \leqslant k$ and $x_1 \cup \ldots \cup x_k$ ($x_1 \cap \ldots \cap x_k$) is the greatest (least) element in K. Note that K is closed under $\cap$ and $\cup$, i.e.,

$$x, y \in K \quad \Rightarrow \quad x \cap y, \ x \cup y \in K.$$

Therefore, it follows from the definition of generating set that $A = K$ and hence $\langle A, \cup, \cap \rangle$ contains the greatest and the least element. □

It can be shown by means of an example that finitely generated lattices (non-distributive) need not be finite. It is also known that

Lemma 1.14. *Each finitely generated distributive lattice is finite.*

Proof. The proof is by induction on the number of generators. First, let us observe that a lattice generated by one element is degenerate (i.e., one-element). Then, suppose that the distributive lattice $\langle A, \cup, \cap \rangle$ has $k + 1$ generators $a_1, \ldots, a_{k+1}$ and that the theorem holds for distributive lattices with k generators.

It follows from the above lemma that A contains the unit 1 and the zero 0. By the inductive hypothesis the sublattice $\langle K, \cup, \cap \rangle$ of $\langle A, \cup, \cap \rangle$ generated by the set $\{a_1, \ldots, a_k\}$ is finite. Now we shall consider the set

$$B = \{x \in A : x = (c \cap a_{k+1}) \cup d \quad \text{for some} \quad c, d \in K \cup \{0, 1\}\}.$$

Let us prove

$$x, y \in B \quad \Rightarrow \quad x \cup y, \ x \cap y \in B.$$

Suppose that $x = (c_1 \cap a_{k+1}) \cup d_1$ and $y = (c_2 \cap a_{k+1}) \cup d_2$ for some $c_1, c_2, d_1, d_2 \in K \cup \{0, 1\}$. Then

$$x \cap y = ((c_1 \cap a_{k+1}) \cup d_1) \cap ((c_2 \cap a_{k+1}) \cup d_2)$$
$$= (c_1 \cup d_1) \cap (a_{k+1} \cup d_1) \cap (c_2 \cup d_2) \cap (a_{k+1} \cup d_2)$$
$$= (c_1 \cup d_1) \cap (c_2 \cup d_2) \cap (a_{k+1} \cup (d_1 \cap d_2))$$
$$= ((c_1 \cup d_1) \cap (c_2 \cup d_2) \cap a_{k+1}) \cup (d_1 \cap d_2).$$

But $\langle K, \cup, \cap \rangle$ is a subalgebra of $\langle A, \cup, \cap \rangle$, hence $K \cup \{0, 1\}$ is closed under the operations $\cap$ and $\cup$. Thus, $x \cap y \in B$ for every $x, y \in B$. We also have $x \cup y \in B$ as

$$x \cup y = ((c_1 \cap a_{k+1}) \cup d_1) \cup ((c_2 \cap a_{k+1}) \cup d_2) = ((c_1 \cup c_2) \cap a_{k+1}) \cup (d_1 \cup d_2).$$

Obviously, B is finite and $a_i \in B$ for each $i \leqslant k + 1$. Since $\{a_1, \ldots, a_{k+1}\}$ is a set of generators of $\langle A, \cup, \cap \rangle$, we get $A = B$ which proves that A is finite. □

It is worth noticing that the same theorem holds for Boolean algebras but it is false for Heyting algebras. We can construct an infinite Heyting algebra $\langle A, \dot\rightarrow, \cup, \cap, - \rangle$ generated by a single element only. Namely — following L. Rieger [109], 1949 — let

$$A = \{2^i \cdot 3^j : -1 \leqslant i - j \leqslant 2\} \cup \{\omega\}$$

and let the lattice ordering $\leqslant_A$ be given by the relation of divisibility (assuming that each natural number divides ω and ω divides no natural number). It can be shown that $\leqslant_A$ is a Heyting ordering on A and that the algebra $\langle A, \dot\rightarrow, \cup, \cap, - \rangle$ is generated by $2^1 \cdot 3^0$.

Filters

Assume that $\langle A, \leqslant \rangle$ is a lattice ordered set and let $\emptyset \neq H \subseteq A$. Then H is called a *filter* in A iff the following holds for each $x, y \in A$:

(i) $x, y \in H \quad \Rightarrow \quad x \cap y \in H$,

(ii) $x \leqslant y \land x \in H \quad \Rightarrow \quad y \in H$.

The filter A in the lattice $\langle A, \cup, \cap \rangle$ is called the *improper* one and $\{x \in A : a \leqslant x\}$ is said to be the *principal filter* determined by the element $a \in A$. We can show the following characterization of filters in implicative lattices (see [106], 1974).

Lemma 1.15. *Let $H \subseteq A$ and let $\mathscr{A} = \langle A, \dot\rightarrow, \cup, \cap \rangle$ be an implicative lattice. Then H is a filter in $\mathscr{A}$ iff*

(i) $1 \in H$;

(ii) $x , \; x \dot\rightarrow y \in H \; \Rightarrow \; y \in H, \quad$ *for each $x, y \in A$.*

Let $\mathscr{A}$ be an implicative lattice (Heyting or Boolean algebra). Assume that $x \dot\leftrightarrow y$ is the abbreviation for $(x \dot\rightarrow y) \cap (y \dot\rightarrow x)$ where $x, y \in A$. Each filter H in $\mathscr{A}$ determines the binary relation $\sim_H$ on A by

$$x \sim_H y \qquad \text{iff} \qquad x \dot\leftrightarrow y \in H.$$

The relation $\sim_H$ (which will be noted in short as $\sim$ when the filter H is fixed) is a congruence on the algebra $\mathscr{A}$. The family of abstraction classes $A/\sim_H$ will be denoted by A/H. In A/H we define the relation $\leqslant_H$:

$$[x] \leqslant_H [y] \qquad \text{iff} \qquad x \dot\rightarrow y \in H$$

which (as can be easily seen) is a lattice ordering in A/H such that:

$$[x] = 1 \Leftrightarrow x \in H$$
$$[x \cap y] = [x] \cap [y]$$
$$[x \cup y] = [x] \cup [y]$$
$$[x \dot\rightarrow y] = [x] \dot\rightarrow [y]$$
$$([-x] = -[x]).$$

The quotient algebra will be noted as $\mathscr{A}/H$. Note that $\mathscr{A}/H$ is not degenerate if H is a proper filter. It is clear that the canonical mapping

$$h(x) = [x]_H$$

is a homomorphism from $\mathscr{A}$ onto $\mathscr{A}/H$.

Theorem 1.16. *Let $\mathscr{A}$, $\mathscr{B}$ be two implicative lattices (Heyting or Boolean algebras) and let $h : A \to B$ be a homomorphism from $\mathscr{A}$ onto $\mathscr{B}$. Then, for every filter H in $\mathscr{B}$, the set $h^{-1}(H)$ is a filter in $\mathscr{A}$ and*

$$\mathscr{A}/h^{-1}(H) \cong \mathscr{B}/H.$$

Proof. Of course, $h^{-1}(H)$ is a filter in $\mathscr{A}$. Further we have

$$x \dot{\to} y \in h^{-1}(H) \quad \Leftrightarrow \quad h(x) \dot{\to} h(y) \in H,$$
$$x \sim y \quad \Leftrightarrow \quad h(x) \sim h(y)$$

for every $x, y \in A$. Hence the mapping g defined by

$$g([x]) = [h(x)]$$

is one-to-one and maps the algebra $\mathscr{A}/h^{-1}(H)$ onto $\mathscr{B}/H$. Moreover, the following condition is fulfilled,

$$[x] \leqslant [y] \quad \Leftrightarrow \quad g([x]) \leqslant g([y]).$$

Thus, by Lemma 1.12, the algebras $\mathscr{A}/h^{-1}(H)$ and $\mathscr{B}/H$ are isomorphic. $\qquad\square$

In particular, if $h : \mathscr{A} \to \mathscr{B}$ maps $\mathscr{A}$ onto $\mathscr{B}$, then $\mathscr{A}/h^{-1}(\{1\}) \cong \mathscr{B}$.

Corollary 1.17. *Let $\mathscr{A}$ be a non-degenerate Heyting algebra. Then*

(i) *the set $G(\mathscr{A})$ of all dense elements is a proper filter in $\mathscr{A}$,*

(ii) *if H is a filter in $\mathscr{A}$ and if $G(\mathscr{A}) \subseteq H$, then $\mathscr{A}/H$ is a Boolean algebra.*

Proof. Trivial by Lemma 1.11 (ii) and (iii). $\qquad\square$.

The family of all proper filters of the lattice $\langle A, \cap, \cup \rangle$ is ordered by the relation of inclusion. A maximal (with respect to the inclusion) proper filter is called an *ultrafilter*. Assume that A contains the least element 0. We say that $X \subseteq A$ has the *finite intersection property* iff $\inf(Y) \neq 0$ for every $Y \in \text{Fin}^*(X)$, i.e.,

$$a_1 \cap \ldots \cap a_k \neq 0 ; \quad \text{for every } a_1, \ldots, a_k \in X.$$

It is easy to prove that X has the finite intersection property iff X is contained in some proper filter. The basic theorem concerning filters in lattices is the following equivalent of the Axiom of Choice.

Theorem 1.18. *If a lattice $\mathscr{A} = \langle A, \cap, \cup \rangle$ contains the zero-element and if $X \subseteq A$ has the finite intersection property, then there exists an ultrafilter H such that $X \subseteq H$.*

The proof based on Zorn's lemma is simple and generally known.

A proper filter H fulfilling, for all $x, y \in A$, the condition

$$x \cup y \in H \quad \Rightarrow \quad x \in H \ \vee \ y \in H$$

is said to be *prime*. Not all prime filters are ultrafilters. In the same way, not all ultrafilters are prime. It can be shown, however, that if the lattice is complementary, then each prime filter is an ultrafilter.

We will also say that a lattice $\mathscr{A}$ is *prime* iff

$$x \cup y = 1_A \quad \Rightarrow \quad x = 1_A \ \vee \ y = 1_A.$$

The connection between prime filters and algebras is stated by

Theorem 1.19. *If H is a proper filter in a lattice $\mathscr{A} = \langle A, \cap, \cup \rangle$, then the following conditions are equivalent:*

(i)   *H is a prime filter;*

(ii) *$\mathscr{A}/H$ is a prime lattice.*

Note that a Boolean algebra is prime if and only if it contains at most two elements. Hence, on the basis of Theorem 1.19, we can easily deduce that each prime filter in a Boolean algebra is an ultrafilter and conversely that each ultrafilter is prime.

Theorem 1.20. *Let H be a proper filter in a Heyting algebra $\mathscr{A}$. Then, for every $x \notin H$ there is a prime filter H_1 such that $x \notin H_1$ and $H \subseteq H_1$.*

It is evident that each proper filter in $\mathscr{A}$ is contained in some ultrafilter (see Theorem 1.18) and that each non-zero element belongs to some ultrafilter. Hence if x is not dense, then there exists an ultrafilter H such that $-x \in H$, i.e., $x \notin H$. On the other hand, it follows from Theorem 1.20 that for each non-unit element x there exists a prime filter H such that $x \notin H$.

Theorem 1.21. *If H is a proper filter in a Heyting algebra $\mathscr{A}$, then the following conditions are equivalent:*

(i)   *H is an ultrafilter;*

(ii)   *$x \in H \ \vee \ -x \in H$, for each $x \in A$;*

(iii) *$G(\mathscr{A}) \subseteq H$ and H is prime;*

(iv)   *$\mathscr{A}/H$ is a two-element Boolean algebra.*

Proof. (i)⇒(ii): If $x \notin H$ and H is an ultrafilter, then $H \cup \{x\}$ does not have the finite intersection property. Thus, $x \cap y = 0$ for some $y \in H$ and hence $-x \in H$.

(ii)⇒(iii): $G(\mathscr{A}) \subseteq H$ by (ii) and Lemma 1.11 (iii). Suppose that H is not prime, i.e., $x \cup y \in H$ for some $x \notin H$ and $y \notin H$. Then, on the basis of (ii), $-x$ and $-y$ are in H, so $0 = (x \cup y) \cap -x \cap -y \in H$, which is impossible.

(iii)⇒(iv): By Corollary 1.17 (ii), $\mathscr{A}/H$ is a non-degenerate Boolean algebra. Moreover, $\mathscr{A}/H$ is prime by Theorem 1.19. Thus, $\mathscr{A}/H$ must be a two-element algebra.

(iv)⇒(i): This is obvious since $x \sim_H 0$ for any $x \notin H$. $\qquad\square$

It is then clear that in Heyting algebras each ultrafilter is a prime filter. The reverse implication, however, is not true. Since in any Boolean algebra the unit is the only dense element, we obtain by Theorem 1.21 (cf. [107], 1963)

Corollary 1.22. *If H is a proper filter in a Boolean algebra $\mathscr{A}$, then the following conditions are equivalent:*

(i)   *H is an ultrafilter;*

(ii) *H is prime;*

(iii) *$x \in H \ \lor \ -x \in H$,   for each $x \in A$;*

(iv) *$\mathscr{A}/H$ is two-element.*

The operation of addition

Constructions of new algebras from given ones play a very important role in algebraic considerations. Two methods of constructing new algebras (lattices) have been already discussed: namely the construction of subalgebras and that of forming quotient algebras. Some further methods are discussed below.

Let $\mathscr{A}_1 = \langle A_1, \dot{\rightarrow}_1, \cup_1, \cap_1, -_1 \rangle$ be a Heyting algebra and $\mathscr{A}_2 = \langle A_2, \dot{\rightarrow}_2, \cup_2, \cap_2 \rangle$ be an implicative lattice. The order relations in $\mathscr{A}_1$ and $\mathscr{A}_2$ are denoted by $\leqslant_1$ and $\leqslant_2$, respectively. We assume that $A_1 \cap A_2 = \emptyset$. The operations $\dot{\rightarrow}, \cup, \cap, -$ are determined on the set $A_1 \cup A_2$ in the following way:

$$
x \cup y = \begin{cases} x \cup_i y & \text{if } x, y \in A_i \\ y & \text{if } x \in A_1 \text{ and } y \in A_2 \\ x & \text{otherwise,} \end{cases}
$$

$$
x \cap y = \begin{cases} x \cap_i y & \text{if } x, y \in A_i \\ x & \text{if } x \in A_1 \text{ and } y \in A_2 \\ y & \text{otherwise,} \end{cases}
$$

$$x \dot{\rightarrow} y = \begin{cases} y & \text{if } x \in A_2 \text{ and } y \in A_1 \\ x \dot{\rightarrow}_i y & \text{if } x, y \in A_i \text{ and } x \dot{\rightarrow}_i y \neq 1_{A_i} \\ 1_{A_2} & \text{otherwise,} \end{cases}$$

$$-x = \begin{cases} -_1 x & \text{if } x \in A_1 \text{ and } x \neq 0_{A_1} \\ 0_{A_1} & \text{if } x \in A_2 \\ 1_{A_2} & \text{if } x = 0_{A_1}. \end{cases}$$

The algebra $\langle A_1 \cup A_2, \dot{\rightarrow}, \cup, \cap, - \rangle$ will be denoted by $\mathscr{A}_1 \oplus \mathscr{A}_2$.

Theorem 1.23. *If $\mathscr{A}_1$ is a Heyting algebra, $\mathscr{A}_2$ is an implicative lattice and $A_1 \cap A_2 = \emptyset$, then $\mathscr{A}_1 \oplus \mathscr{A}_2$ is a Heyting algebra with the order relation defined by*

$$x \leqslant y \quad \Leftrightarrow \quad (x \in A_1 \wedge y \in A_2) \vee x \leqslant_1 y \vee x \leqslant_2 y.$$

The operation $\oplus$ is a generalization of *Jaśkowski's mast operation* (cf. [44], 1936; [127], 1965; [150], 1974). There was defined a more general *operation of gluing* for lattices by Herrmann [41], 1973, which was developed and employed by Grygiel [36], 2004, to analyze the structure of distributive lattices. We will use the operation $\oplus$ only when one of the algebras is degenerate (i.e., is one-element). For these special cases a special notation will be used:

$$\mathscr{A} \oplus = \mathscr{A} \oplus \mathscr{A}_2, \text{ when } \mathscr{A}_2 \text{ is degenerate;}$$
$$\oplus \mathscr{A} = \mathscr{A}_1 \oplus \mathscr{A}, \text{ when } \mathscr{A}_1 \text{ is degenerate.}$$

It can be said that $\mathscr{A} \oplus$ (or $\oplus \mathscr{A}$) is obtained from $\mathscr{A}$ by the addition of the greatest (or the least) element. Thus, each implicative lattice $\mathscr{A}$ can be extended to the Heyting algebra $\oplus \mathscr{A}$. Note that $\mathscr{A}$ is a substructure of $\oplus \mathscr{A}$, i.e., the operations in $\mathscr{A}$ are the restrictions of the operations from $\oplus \mathscr{A}$. We also get

Lemma 1.24. *For all Heyting algebras $\mathscr{A}$ and $\mathscr{B}$:*

(i) *if $\mathscr{A}$ is embeddable in $\mathscr{B}$, then $\mathscr{A} \oplus$ is embeddable in $\mathscr{B} \oplus$;*

(ii) *$\mathscr{A} \oplus$ is a prime algebra.*

It is also easy to verify that $\mathscr{A} \cong \mathscr{A} \oplus / H$ where $H = \{1_A, 1_{A \oplus}\}$.

Lemma 1.25. *Let $\mathscr{A}$ be a Heyting algebra such that $A \setminus \{1_A\}$ contains the greatest element y. Then $H = \{y, 1_A\}$ is a filter in $\mathscr{A}$ and $(\mathscr{A}/H) \oplus \cong \mathscr{A}$.*

Proof. Of course, H is the principal filter determined by the element y. Let us prove that

$$a \sim_H b \quad \Leftrightarrow \quad a = b \vee a, b \in H, \quad \text{for each } a, b \in A.$$

The implication ($\Leftarrow$) is obvious. Suppose, on the other hand, that $a \sim_H b$ for some $a \notin H$ and $b \notin H$. Then, according to the definition of $\sim_H$, we have $y \leqslant a \dot{\rightarrow} b$ and

$y \leqslant b \rightarrow a$. Hence, $y \cap a \leqslant b$ and $y \cap b \leqslant a$. But y is the greatest element in $A \setminus \{1_A\}$ and $a \neq 1_A$, $b \neq 1_A$. Thus, $y \cap a = a$ and $y \cap b = b$. Therefore $a \leqslant b$ and $b \leqslant a$ which means that $a = b$.

If we assume that $a \sim_H b$ and $a \in H$ (or $b \in H$), then also $b \in H$ (or $a \in H$) and this completes the proof of the above equivalence.

Thus, $[a]_H = \{a\}$ for all $a \in A \setminus H$ and hence the mapping f defined by

$$f(a) = \begin{cases} [a]_H & \text{if } a \neq 1_A \\ 1 & \text{if } a = 1_A \end{cases}$$

where 1 is the greatest element in $(\mathscr{A}/H)\oplus$, is one-to-one, maps the algebra $\mathscr{A}$ onto $(\mathscr{A}/H)\oplus$ and fulfills the condition

$$a \leqslant b \quad \Leftrightarrow \quad f(a) \leqslant f(b).$$

Consequently, by Lemma 1.12, the algebras $\mathscr{A}$ and $(\mathscr{A}/H)\oplus$ are isomorphic. $\square$

Let $\mathscr{A}_t$, for $t \in T$, be an implicative lattice with $\leqslant_t$ as the order relation. The relation $\leqslant$ in the product $\mathbf{P}_{t\in T}\, A_t$ is determined as

$$\langle x_t \rangle_{t\in T} \leqslant \langle y_t \rangle_{t\in T} \quad \Leftrightarrow \quad x_t \leqslant_t y_t \text{ for all } t \in T.$$

The algebra $\mathbf{P}_{t\in T}\, A_t$ is an implicative lattice with $\leqslant$ as the lattice ordering. It is also evident that $\langle 1_{A_t} \rangle_{t\in T}$ is the greatest element in the product and that $\langle 0_{A_t} \rangle_{t\in T}$ is the least element (provided that 0_A occurs in each A_t). Finally, when $\mathscr{A}_t$ are Heyting algebras, then $\mathbf{P}_{t\in T}\, \mathscr{A}_t$ is a Heyting algebra and the product is a Boolean algebra if $\{\mathscr{A}_t\}_{t\in T}$ is an indexed family of Boolean algebras.

Jaśkowski algebras

Product and addition operations allow us to construct some family of Heyting algebras which is (as we shall see later) adequate for intuitionistic propositional logic. This completeness theorem was first proved by S. Jaśkowski [44], 1936.

Let us define the class of *Jaśkowski algebras* as the least class of Heyting algebras containing any degenerate algebra and closed with respect to the operation of addition and finite products. In other words:

a. Degenerate algebras are Jaśkowski algebras.

b. If $\mathscr{A}$ is a Jaśkowski algebra, then so is $\mathscr{A}\oplus$.

c. If $\mathscr{A}_1, \ldots, \mathscr{A}_k$ belong to the considered class of algebras, then $\mathscr{A}_1 \times \ldots \times \mathscr{A}_k$ is also a Jaśkowski algebra.

From the class of Jaśkowski algebras we choose the sequence $\langle \mathscr{J}_n \rangle_{n \geqslant 1}$ as follows: let $\mathscr{J}_1$ be any degenerate algebra and let

$$\mathscr{J}_{n+1} = (\mathscr{J}_n)^n \oplus \quad \text{for each } n \geqslant 1.$$

Lemma 1.26. *The algebra $\mathscr{J}_n$ is embeddable in $\mathscr{J}_{n+1}$, for each $n \geqslant 2$.*

Proof. By induction on n. First, we observe that $\mathscr{J}_2$ is embeddable in each non-degenerate Heyting algebra. We assume next that $\mathscr{J}_n$ is embeddable in $\mathscr{J}_{n+1}$. Then, by Corollary 1.8, $(\mathscr{J}_n)^n$ is embeddable in $(\mathscr{J}_{n+1})^n$. Moreover, by Lemma 1.9 (iii), $(\mathscr{J}_{n+1})^n$ is embeddable in $(\mathscr{J}_{n+1})^{n+1}$, hence $(\mathscr{J}_n)^n$ is embeddable in $(\mathscr{J}_{n+1})^{n+1}$. Using Lemma 1.24 (i) we conclude that there is an embedding of $\mathscr{J}_{n+1}$ into $\mathscr{J}_{n+2}$. $\qquad\square$

Corollary 1.27. *If $1 < n \leqslant m$, then $\mathscr{J}_n$ is embeddable in $\mathscr{J}_m$.*

Lemma 1.28. *For each Jaśkowski algebra $\mathscr{A}$ there are integers $n, m \geqslant 1$ such that $\mathscr{A}$ is embeddable in $(\mathscr{J}_n)^m$.*

Proof. Any Heyting algebra is a Jaśkowski algebra if it can be constructed from a degenerate algebra by means of the operation of addition and finite products.

  a. Obviously, the statement is true for degenerate algebras.

  b. Suppose that $\mathscr{A}$ is embeddable in $(\mathscr{J}_n)^m$. If $n = 1$, then $(\mathscr{J}_n)^m$ is degenerate and hence $\mathscr{A}$ is degenerate, too; which implies $\mathscr{A} \oplus \cong \mathscr{J}_2$.
  Assume that $n > 1$ and let $k = \max\{n, m\}$. It follows from Corollary 1.27 that $\mathscr{J}_n$ is embeddable in $\mathscr{J}_k$ and hence, by Corollary 1.8 and Lemma 1.9 (iii), $(\mathscr{J}_n)^m$ is embeddable in $(\mathscr{J}_k)^k$. Thus, $\mathscr{A}$ is embeddable in $(\mathscr{J}_k)^k$ which implies, on the basis of Lemma 1.24, the existence of an embedding from $\mathscr{A} \oplus$ into $\mathscr{J}_{k+1}$.

  c. Suppose that $\mathscr{A}_i$, for $1 \leqslant i \leqslant k$ is embeddable in $(\mathscr{J}_{n_i})^{m_i}$. If all algebras $\mathscr{A}_i$ are degenerate, then the product $\mathbf{P}_{i \leqslant k} \mathscr{A}_i$ is also degenerate and hence the product is embeddable in $\mathscr{J}_1$. Thus, we claim that the set

$$I = \{i \leqslant k : \mathscr{A}_i \text{ is not degenerate}\}$$

  is not empty. Observe that $n_i > 1$ if $i \in I$. Let $n = \max\{n_i : i \in I\}$. By Corollary 1.27, $\mathscr{J}_{n_i}$ is embeddable in $\mathscr{J}_n$ for all $i \in I$. Hence it follows from Corollary 1.8 and from our assumption that $\mathscr{A}_i$ is embeddable in $(\mathscr{J}_n)^{m_i}$ for all $i \in I$. Thus, by Corollary 1.8, the product $\mathbf{P}_{i \in I} \mathscr{A}_i$ is embeddable in $\mathbf{P}_{i \in I}(\mathscr{J}_n)^{m_i}$. As (by Lemma 1.9 (iv), (i))

$$\mathbf{P}_{i \leqslant k} \mathscr{A}_i \cong \mathbf{P}_{i \in I} \mathscr{A}_i \quad \text{and} \quad \mathbf{P}_{i \in I}(\mathscr{J}_n)^{m_i} \cong (\mathscr{J}_n)^m$$

  where $m = \sum_{i \in I} m_i$, we conclude $\mathbf{P}_{i \leqslant k} \mathscr{A}_i$ is embeddable in $(\mathscr{J}_n)^m$. $\quad\square$

The next lemma paves the way to the mentioned completeness theorem for the intuitionistic propositional logic.

Lemma 1.29. *Each finite Heyting algebra is embeddable in some Jaśkowski algebra.*

Proof. It is obvious that the theorem holds for any degenerate Heyting algebra. Assume that the theorem is true for all Heyting algebras with less than $k + 1$ elements and suppose that $\mathscr{A} = \langle A, \rightarrowtail, \cup, \cap, - \rangle$ has exactly $k + 1$ elements.

Let Y be the set of all maximal elements in $A \setminus \{1_A\}$. It is obvious that Y is non-empty since A is finite. We have to consider the following cases:

(1) Y is one-element, i.e., $Y = \{y\}$ for some $y \in A$,

(2) Y contains (at least) two elements.

If (1) happens, $H = \{1_A, y\}$ is the principal filter determined by y. By the inductive hypothesis $\mathscr{A}/H$ is embeddable in some Jaśkowski algebra $\mathscr{J}$. Hence $(\mathscr{A}/H)\oplus$ is embeddable in $\mathscr{J}\oplus$ (see 1.24 (i)) and, by 1.25, the algebras $\mathscr{A}$ and $(\mathscr{A}/H)\oplus$ are isomorphic. Then $\mathscr{A}$ is embeddable in the Jaśkowski algebra $\mathscr{J}\oplus$.

Assume that Y contains (at least) two elements. The algebra $\mathscr{A}/H_y$, where $H_y = \{1_A, y\}$ is the principal filter generated by an element y of Y, contains less than $k + 1$ elements. Then, by the inductive hypothesis, for every $y \in Y$ there exists a Jaśkowski algebra $\mathscr{J}_y$ and a one-to-one homomorphism f_y from $\mathscr{A}/H_y$ into $\mathscr{J}_y$. Let $h_y : A \to A/H_y$ be the canonical mapping. Since the composition $f_y \circ h_y$ is a homomorphism from $\mathscr{A}$ into $\mathscr{J}_y$, the mapping h defined by

$$h(x) = \langle f_y(h_y(x))\rangle_{y \in Y}, \quad \text{for each } x \in A$$

is a homomorphism from $\mathscr{A}$ into $\mathbf{P}_{y \in Y}\,\mathscr{J}_y$ — see Lemma 1.6.

The product $\mathbf{P}_{y \in Y}\,\mathscr{J}_y$ will be denoted then by $\mathscr{J}$. We shall now prove that $h(y) \neq 1_J$ for all $y \in Y$.

Suppose, on the contrary that $h(x) = 1_J$ for some $x \in Y$. Then we have $f_y(h_y(x)) = 1_{J_y}$, for every $y \in Y$. Since f_y is one-to-one, $h_y(x)$ is the unit-element in $\mathscr{A}/H_y$, that is $x \in H_y = \{1_A, y\}$. We get $y = x$ for each $y \in Y$. Hence Y is one-element, which contradicts our assumptions.

Let us prove that h is an embedding. Suppose $a \neq b$, i.e., we can assume that $a \rightarrowtail b \neq 1_A$ (if $b \rightarrowtail a \neq 1_A$, the proof is similar). Then there is an element $y \in Y$ such that $a \rightarrowtail b \leqslant y$, hence $h(a) \rightarrowtail h(b) = h(a \rightarrowtail b) \leqslant h(y) \neq 1_J$, which means that $h(a) \neq h(b)$.

It has been shown that h is an embedding from $\mathscr{A}$ into $\mathscr{J}$. Since $\mathscr{J}$ is a finite product of Jaśkowski algebras $\mathscr{J}_y$, we conclude that $\mathscr{A}$ is embeddable in some Jaśkowski algebra. $\square$

Given a formula α which is not valid in a finite Heyting algebra $\mathscr{A}$, we can find a Jaśkowski algebra $\mathscr{J}_n$ such that α is not valid in $\mathscr{J}_n$, either, using the following corollary of Lemmas 1.28 and 1.29:

Corollary 1.30. *For every finite Heyting algebra $\mathscr{A}$ there are integers $n, m \geqslant 1$ such that $\mathscr{A}$ is embeddable in $(\mathscr{J}_n)^m$.*

Linear algebras

Let $\langle A, \leqslant \rangle$ be a linearly ordered set, i.e., let $\leqslant$ be an ordering on A such that

$$x \leqslant y \ \lor \ y \leqslant x, \qquad \text{for each } x, y \in A.$$

Clearly, $\leqslant$ is a lattice ordering with the bounds defined by

$$x \cup y = \max\{x, y\}, \quad x \cap y = \min\{x, y\}.$$

If we assume that $\langle A, \leqslant \rangle$ contains the unit-element, then the relative pseudo-complement $x \dot{\to} y$ will exist for every $x, y \in A$

$$x \dot{\to} y = \begin{cases} 1_A & \text{if } x \leqslant y \\[2mm] y & \text{if } y < x. \end{cases}$$

Hence, any linearly ordered set with the unit- and the zero-element can be considered as a prime Heyting algebra. Such algebras will be said to be *linear*. It should be noticed that each finite linearly ordered set contains the unit- and the zero-element and then it can be considered as a linear Heyting algebra.

Lemma 1.31. *If $\mathscr{A}$, $\mathscr{B}$ are non-degenerate linear Heyting algebras, $\mathscr{A}$ is finite and $Nc(A) \leqslant Nc(B)$, then $\mathscr{A}$ is embeddable in $\mathscr{B}$.*

Proof. In order to prove this lemma it suffices to observe that any one-to-one mapping from A into B which preserves the ordering relation, the unit- and the zero-element (i.e., $h(1_A) = 1_B$, $h(0_A) = 0_B$ and $x \leqslant y \Rightarrow h(x) \leqslant h(y)$) is an embedding from $\mathscr{A}$ into $\mathscr{B}$. The existence of such a mapping is quite obvious whenever A is finite and $2 \leqslant Nc(A) \leqslant Nc(B)$. $\square$

It is easy to see that any subalgebra and any quotient algebra of a linear Heyting algebra is also linearly ordered. Moreover, each subset of a linear algebra containing the zero- and the unit-element is closed under its operations $\dot{\to}$, $\cup$, $\cap$, $-$ (i.e., forms a subalgebra).

Let us define the sequence $\langle \mathscr{G}_n \rangle_{n \geqslant 1}$ of *Gödel–Heyting algebras*. We choose this sequence from the class of Jaśkowski algebras. Namely, let $\mathscr{G}_1$ be a degenerate algebra and $\mathscr{G}_{n+1} = \mathscr{G}_n \oplus \mathscr{G}_1$. Note that the algebras $\mathscr{G}_n$ are finite and linearly ordered. We will also consider the infinite linear algebra $\mathscr{G}_\infty$ which is determined by ordering of real numbers on the set

$$\left\{ \frac{1}{n} : n \text{ is a natural number} \right\} \cup \{0\}.$$

Corollary 1.32.

(i) *If $1 < n \leqslant m$, then $\mathscr{G}_n$ is embeddable in $\mathscr{G}_m$;*

(ii) *$\mathscr{G}_n$ is embeddable in $\mathscr{G}_\infty$ for each $n > 1$;*

(iii) *Each finite linear algebra $\mathscr{A}$ is isomorphic with the algebra $\mathscr{G}_n$, where n is the cardinality of A.*

This follows directly from Lemma 1.31.

1.4 Propositional logics

Let $\mathscr{S} = \langle S, F_1, \ldots, F_n \rangle$ be the algebra of a fixed propositional language. Any pair $\langle R, X \rangle$, where R is a set of rules (i.e., $R \subseteq \mathscr{R}_S$) and $X \subseteq S$, is called a *system of propositional logic* (or a *propositional logic*). We say that X is the set of *axioms* (of this logic) and R is the set of its *primitive rules*.

Consequence operations

A mapping $Cn : 2^S \to 2^S$ is called *a consequence operation* over S — or a *closure operation* — iff

(i) $X \subseteq Cn(X)$,

(ii) $X \subseteq Y \;\Rightarrow\; Cn(X) \subseteq Cn(Y)$,

(iii) $Cn(Cn(X)) \subseteq Cn(X)$,   for each $X, Y \subseteq S$.

A consequence Cn is said to be *finitistic* iff

(iv) $Cn(X) = \bigcup\{Cn(Y) : Y \in \mathrm{Fin}(X)\}$,   for each $X \subseteq S$

and it is *compact* if for each $Y \subseteq S$ there is an $X \in \mathrm{Fin}(Y)$ such that

(v) $Cn(Y) = S \;\Rightarrow\; Cn(X) = S$.

Let Cn be a finitistic consequence operation. Then, one easily proves that Cn is compact if and only if $Cn(\{\alpha_1, \ldots, \alpha_k\}) = S$ for some $\alpha_1, \ldots, \alpha_k \in S$.

A trivial example of a consequence operation is the inconsistent operation, i.e., the consequence Cn such that $Cn(X) = S$ for every $X \subseteq S$. A consequence Cn is said to be *consistent* ($Cn \in \mathrm{CNS}$) iff

(vi) $Cn(\emptyset) \neq S$

and Cn is *complete (Post-complete)* iff

(vii) $Cn(\{\alpha\}) = S$, for each $\alpha \notin Cn(\emptyset)$.

Let us define an ordering relation $\leqslant$ on the family of all consequence operations over S. If Cn_1, Cn_2 are two given consequences, then

$$Cn_1 \leqslant Cn_2 \quad \Leftrightarrow \quad (Cn_1(X) \subseteq Cn_2(X)\,, \quad \text{for each } X \subseteq S).$$

It is evident that

Lemma 1.33. *The following conditions are equivalent:*

(i) $Cn_1 \leqslant Cn_2$;

(ii) $Cn_2 \circ Cn_1 = Cn_2$;

(iii) $Cn_1 \circ Cn_2 = Cn_2$.

Given a family $\{Cn_t : t \in T\}$ of consequences, we define the operations $\prod_{t\in T} Cn_t$, $\coprod_{t\in T} Cn_t$ as follows:

Definition 1.34.

(i) $\left(\prod_{t\in T} Cn_t\right)(X) = \bigcap\{Cn_t(X) : t \in T\}$;

(ii) $\left(\coprod_{t\in T} Cn_t\right)(X) = \bigcap\{Cn(X) : Cn_t \leqslant Cn \text{ for all } t \in T\}$,   for $X \subseteq S$.

It is easy to prove that both $\prod_{t\in T} Cn_t$ and $\coprod_{t\in T} Cn_t$ are consequence operations over S and, what is more,

$$\left(\coprod_{t\in T} Cn_t\right)(X) = \bigcap\{Y : X \subseteq Y = Cn_t(Y) \text{ for all } t \in T\}.$$

Lemma 1.35. *The family of all consequences over S is a complete lattice with $\leqslant$ as the lattice-ordering and*

$$\inf\{Cn_t : t \in T\} = \prod_{t\in T} Cn_t \quad , \quad \sup\{Cn_t : t \in T\} = \coprod_{t\in T} Cn_t.$$

The greatest element of the lattice is the inconsistent operation and the operation Id such that $Id(X) = X$ for all $X \subseteq S$ is the least consequence over S. It should be noted that the operator $\prod$ does not preserve the finiteness of the consequences (in contrast to $\coprod$).

Let Cn be a consequence operation over S. A set $X \subseteq S$ is said to be Cn-closed iff $X = Cn(X)$. The family of all Cn-closed sets is closed under arbitrary intersections, i.e.,

$$Cn\left(\bigcap\{Cn(X_t) : t \in T\}\right) = \bigcap\{Cn(X_t) : t \in T\}$$

for each $\{X_t : t \in T\} \subseteq 2^S$. This family need not be, however, closed under unions. We can only prove

Lemma 1.36. *If Cn is finitistic and if $\{X_t : t \in T\}$ is a chain of sets, then*

$$Cn\left(\bigcup\{X_t : t \in T\}\right) = \bigcup\{Cn(X_t) : t \in T\}.$$

We conclude that the family of all Cn-closed sets is a closure (or absolutely multiplicative) system; the relation of inclusion is a lattice ordering on the family and this lattice is complete with S as the greatest element and $Cn(\emptyset)$ as the least one. Moreover, we have

$$\inf\{X_t : t \in T\} = \bigcap\{X_t : t \in T\},$$
$$\sup\{X_t : t \in T\} = Cn\left(\bigcup\{X_t : t \in T\}\right),$$
$$\text{if } X_t = Cn(X_t) \text{ for all } t \in T.$$

A set $X \subseteq S$ is said to be *Cn-consistent* iff $Cn(X) \neq S$; X is *Cn–maximal* provided that X is Cn-consistent and $Cn(X \cup \{\alpha\}) = S$ for every $\alpha \notin Cn(X)$. If X is Cn-maximal, then $Cn(X)$ is a maximal element in the family of Cn-closed and Cn-consistent sets. A set $X \subseteq S$ is *Cn-axiomatizable* provided that there is a finite set Y such that $Cn(X) = Cn(Y)$.

Lemma 1.37. *If there is a Cn-unaxiomatizable set, then the family of all Cn– axiomatizable and Cn-closed sets is infinite.*

Proof. Assume that X is not Cn-axiomatizable and let us consider the family $\{Cn(Y) : Y \in \mathrm{Fin}(X)\}$. Suppose that the family is finite, i.e.,

$$\{Cn(Y) : Y \in \mathrm{Fin}(X)\} = \{Cn(Y_1), \ldots, Cn(Y_k)\}$$

for some $Y_1, \ldots, Y_k \in \mathrm{Fin}(X)$. Let us take $Y_0 = Y_1 \cup \ldots \cup Y_k$. Then $Y_0 \in \mathrm{Fin}(X)$ and $Y \subseteq Cn(Y_0)$ for every $Y \in \mathrm{Fin}(X)$. Thus, $Cn(X) \subseteq Cn(Y_0)$ which contradicts our assumptions.

It has been shown that $\{Cn(Y) : Y \in \mathrm{Fin}(X)\}$ is infinite. So is the family of all Cn-axiomatizable and Cn-closed sets. $\qquad\square$

Let us observe that the cardinality of all Cn-axiomatizable and Cn-closed sets is less than the cardinality of S, i.e.,

$$Nc\{Cn(X) : X \in \mathrm{Fin}(S)\} \leqslant Nc(S).$$

A set X is said to be *Cn-independent* if $\alpha \notin Cn(X \setminus \{\alpha\})$ for every $\alpha \in X$. Observe that if Cn is finitistic, then any infinite Cn-independent set is not Cn-axiomatizable.

Lemma 1.38. *If X is Cn-independent, then for each $X_1, X_2 \subseteq X$,*

$$X_1 \subseteq X_2 \quad \Leftrightarrow \quad Cn(X_1) \subseteq Cn(X_2).$$

As an immediate result of Lemmas 1.37 and 1.38 we obtain

Corollary 1.39. *If Cn is finitistic over a countable language S and if there is an infinite Cn-independent set, then*

(i) $\mathfrak{c} = Nc\{Cn(X) : X \subseteq S\}$;

(ii) $\aleph_0 = Nc\{Cn(X) : X \in \mathrm{Fin}(S)\}$;

(iii) $\mathfrak{c} = Nc\{Cn(X) : Cn(X) \text{ is not } Cn\text{-axiomatizable}\}$.

Consequences generated by rules of inference

Let R be a set of rules, i.e., let $R \subseteq \mathscr{R}_S$. A set $X \subseteq S$ is said to be *closed under the rules R*, in symbols $R(X)$, provided that for every $r \in R$ and every $\langle \Pi, \alpha \rangle \in r$,

$$\Pi \subseteq X \quad \Rightarrow \quad \alpha \in X.$$

Definition 1.40. For every $X \subseteq S$ and $R \subseteq \mathscr{R}_S$, let

$$Cn(R, X) = \bigcap \{Y \subseteq S : X \subseteq Y \text{ and } R(Y)\}.$$

Then $Cn(R, X)$ can be proved to be the least set containing X and closed with respect to the rules R. We have

Lemma 1.41. *For every* $X \subseteq S$ *and* $R \subseteq \mathscr{R}_S$,

$$Cn(R, X) = X \qquad \Leftrightarrow \qquad R(X).$$

If $\alpha \in Cn(R, X)$, then α is said to be *derivable* from the set X by means of the rules R. It can be shown that a formula α is derivable from X by means of R if and only if α has a *formal proof* on the ground of the system $\langle R, X \rangle$, i.e.,

Lemma 1.42 (fin). $\alpha \in Cn(R, X)$ *iff there is a finite sequence* $\alpha_1, \dots, \alpha_n$ *of formulas such that* $\alpha_n = \alpha$ *and for every* $i \leqslant n$,

$$\alpha_i \in X \ \vee \ \langle \Pi, \alpha_i \rangle \in r \text{ for some } r \in R \text{ and some } \Pi \subseteq \{\alpha_1, \dots, \alpha_{i-1}\}.$$

The set $Cn(R, X)$ can be considered to be a result of operating some mapping Cn on the pair $\langle R, X \rangle$. Each system $\langle R, X \rangle$ determines thus a consequence (finitistic consequence if R is a set of finitary rules) over S:

$$Cn_{RX}(Y) = Cn(R, X \cup Y), \quad \text{for each } Y \subseteq S.$$

We will also write Cn_R if the set X of axioms is empty. It should be noticed that each finitistic consequence operation over S can be generated in this way by some propositional system; i.e., (cf. [64], 1958):

Theorem 1.43. *For each finitistic consequence operation* Cn, *there is a set* $X \subseteq S$ *and a set* $R \subseteq \mathscr{R}_S$ *such that* $Cn = Cn_{RX}$.

Proof. Consider the rule r defined as follows:

$$\langle \Pi, \alpha \rangle \in r \quad \Leftrightarrow \quad \alpha \in Cn(\Pi), \text{ for } \alpha \in S \text{ and } \emptyset \neq \Pi \in \mathrm{Fin}(S).$$

Obviously, $Cn(Y) \subseteq Cn(\{r\}, Cn(\emptyset) \cup Y)$ for every $Y \subseteq S$. As Cn is a consequence,

$$\langle \Pi, \alpha \rangle \in r \wedge \Pi \subseteq Cn(Y) \quad \Rightarrow \quad \alpha \in Cn(Y).$$

Thus, $Cn(Y)$ is r-closed and hence $Cn(\{r\}, Cn(\emptyset) \cup Y) \subseteq Cn(Y)$.   □

Accepting Definition 1.1 (∞) we can prove — in a similar way to the one above — that every consequence Cn is generated by some set $R \subseteq \mathscr{R}_S$ of 'instructions':

Theorem 1.44 (∞). *For each consequence operation* Cn, *there is a set* $R \subseteq \mathscr{R}_S$ *such that* $C = Cn_R$.

If $Cn = Cn_{RX}$, then the system $\langle R, X \rangle$ is said to be a *base* for Cn. It should be noticed, however, that there are plenty of bases for any given consequence operation; the concept of a base is not uniquely defined. On the other hand, each propositional system $\langle R, X \rangle$ determines the consequence operation Cn_{RX} uniquely. This is the reason why we consider propositional logics as pairs $\langle R, X \rangle$ rather than consequence operations.

Obviously, all notions defined in terms of consequence operations can be reformulated in terms of propositional systems, and vice versa. For example, $\langle R, X \rangle$ is said to be *consistent*, $\langle R, X \rangle \in$ Cns, provided that Cn_{RX} is consistent, i.e., $Cn(R, X) \neq S$. The *Post-completeness* of $\langle R, X \rangle$ can be defined now as

$$\langle R, X \rangle \in Cpl \quad \Leftrightarrow \quad \big(Cn(R, X \cup \{\alpha\}) = S, \text{ for every } \alpha \notin Cn(R, X) \big).$$

The notion of compactness, axiomatizability and independency can also be defined in a similar way.

Admissible and derivable rules

For each propositional system $\langle R, X \rangle$, where $R \subseteq \mathscr{R}_S$ and $X \subseteq S$, one can define two sets of rules of inference (see [38], 1965).

Definition 1.45. For each $R \subseteq \mathscr{R}_S$, $r \in \mathscr{R}_S$ and $X \subseteq S$:

(i) $r \in \mathrm{Adm}(R, X) \quad \Leftrightarrow \quad Cn(R \cup \{r\}, X) \subseteq Cn(R, X)$;

(ii) $r \in \mathrm{Der}(R, X) \quad \Leftrightarrow \quad \big(Cn(R \cup \{r\}, X \cup Y) \subseteq Cn(R, X \cup Y) \,, \text{ for all } Y \subseteq S \big).$

The set $\mathrm{Der}(R, X)$ contains all rules *derivable* from R and X, $\mathrm{Adm}(R, X)$ is the set of all rules *admissible* in the system $\langle R, X \rangle$. By Lemma 1.41, we easily get

Lemma 1.46. *For each $R \subseteq \mathscr{R}_S$, $r \in \mathscr{R}_S$ and $X \subseteq S$:*

(i) $r \in \mathrm{Adm}(R, X) \quad \Leftrightarrow \quad r\big(Cn(R, X) \big);$

(ii) $r \in \mathrm{Der}(R, X) \quad \Leftrightarrow \quad \big(r(Cn(R, X \cup Y)) \,, \text{ for all } Y \subseteq S \big).$

Therefore, $r \in \mathrm{Adm}(R, X)$ iff the rule r does not change the set of the formulas provable on the ground of $\langle R, X \rangle$ and $r \in \mathrm{Der}(R, X)$ iff r is admissible in every oversystem of $\langle R, X \rangle$, i.e.,

$$\mathrm{Der}(R, X) = \bigcap \{\mathrm{Adm}(R, X \cup Y) : Y \subseteq S\}.$$

Hence $\mathrm{Der}(R, X) \subseteq \mathrm{Adm}(R, X)$. To show that this inclusion is not reversible, let us consider the following rule over S_2:

$$r : \frac{\alpha \to (\alpha \to \alpha)}{\alpha} \; ; \quad \text{for all } \alpha \in S_2$$

and let $X = \{\alpha \to \alpha : \alpha \in S_2\}$. It is easy to observe that the set X is closed under the rules r and r_0 (modus ponens). Thus, $X = Cn(\{r\}, X) = Cn(\{r_0\}, X) = Cn(\{r, r_0\}, X)$. Hence we get $r \in \text{Adm}(\{r_0\}, X)$ and $r_0 \in \text{Adm}(\{r\}, X)$. But $r \notin \text{Der}(\{r_0\}, X)$ since, for $p \in At$, the set $X_1 = X \cup \{p \to (p \to p)\}$ is closed under r_0 and is not closed with respect to r: $X_1 = Cn(\{r_0\}, X_1) \neq Cn(\{r_0, r\}, X_1)$. Similarly, it can be shown that $r_0 \notin \text{Der}(\{r\}, X)$ because $Cn(\{r_0, r\}, X \cup \{p, p \to q\}) \neq Cn(\{r\}, X \cup \{p, p \to q\}) = X \cup \{p, p \to q\}$ for $p, q \in At$. Thus, $\text{Der}(\{r_0\}, X) \neq \text{Adm}(\{r_0\}, X) = \text{Adm}(\{r\}, X) \neq \text{Der}(\{r\}, X)$ and $\text{Der}(\{r_0\}, X) \neq \text{Der}(\{r\}, X)$.

Some more interesting examples of rules which are $\langle R, X \rangle$ admissible but which are not $\langle R, X \rangle$ derivable will be given later.

Lemma 1.47. *For every $R \subseteq \mathscr{R}_S$, $r \in \mathscr{R}_S$ and $X \subseteq S$:*

(i) $r \in \text{Adm}(R, X) \Leftrightarrow \big([\Pi \subseteq Cn(R, X) \Rightarrow \alpha \in Cn(R, X)], \text{ for each } \langle \Pi, \alpha \rangle \in r\big)$;

(ii) $r \in \text{Der}(R, X) \Leftrightarrow \big(\alpha \in Cn(R, X \cup \Pi), \text{ for each } \langle \Pi, \alpha \rangle \in r\big)$.

Using the above lemma, one can show that $Cn(\text{Adm}(R, X), X) = Cn(R, X)$ and $Cn(\text{Der}(R, X), X \cup Y) = Cn(R, X \cup Y)$. It should be noticed, however, that $Cn(\text{Adm}(R, X), X \cup Y) = Cn(R, X \cup Y)$ need not be true.

The following properties of the operations Adm and Der are immediate from Lemma 1.47

Corollary 1.48. *For every $R \subseteq \mathscr{R}_S$, and $X \subseteq S$:*

(i) $R \subseteq \text{Adm}(R, X)$;

(ii) $\text{Adm}(\text{Adm}(R, X), X) = \text{Adm}(R, X)$.

Let us remark that the operation Adm is not monotonic, see the above example.

Corollary 1.49. *For every R, $R_1 \subseteq \mathscr{R}_S$, and $X, Y \subseteq S$*

(i) $R \subseteq \text{Der}(R, X)$;

(ii) $R \subseteq R_1 \Rightarrow \text{Der}(R, X) \subseteq \text{Der}(R_1, X)$;

(iii) $X \subseteq Y \Rightarrow \text{Der}(R, X) \subseteq \text{Der}(R, Y)$;

(iv) $\text{Der}(\text{Der}(R, X), X) = \text{Der}(R, X)$.

Both the operations Adm and Der are not finitistic, cf. [78], 1969.

Comparison of systems

Let $\langle R, X \rangle$, $\langle R_1, X_1 \rangle$ be two propositional systems over S, i.e., let $R, R_1 \subseteq \mathscr{R}_S$ and $X, X_1 \subseteq S$.

Lemma 1.50. *If X, X_1 are non-empty, then*

(i) $\text{Adm}(R, X) = \text{Adm}(R_1, X_1) \quad \Leftrightarrow \quad Cn(R, X) = Cn(R_1, X_1)$;

(ii) $\mathrm{Der}(R, X) = \mathrm{Der}(R_1, X_1) \quad \Leftrightarrow \quad \forall_{Y \subseteq S} \, Cn(R, X \cup Y) = Cn(R_1, X_1 \cup Y)$.

Proof. (i): The implication ($\Leftarrow$) follows immediately from Lemma 1.46 (i). Now, assume that $\mathrm{Adm}(R, X) = \mathrm{Adm}(R_1, X_1)$ and let $\alpha \in Cn(R, X)$. We have to prove that $\alpha \in Cn(R_1, X_1)$. Suppose that $\beta \in X_1$ and let us consider the one-element rule $r = \{\langle \{\beta\}, \alpha \rangle\}$. Obviously the rule r is admissible for $\langle R, X \rangle$ — see Lemma 1.47 (i) — because $\alpha \in Cn(R, X)$. According to our assumptions $r \in \mathrm{Adm}(R_1, X_1)$. But $\{\beta\} \subseteq X_1 \subseteq Cn(R_1, X_1)$ and hence, by 1.47 (i), $\alpha \in Cn(R_1, X_1)$.

It has been shown that $Cn(R, X) \subseteq Cn(R_1, X_1)$. The proof of the reverse inclusion is quite similar.

(ii): Assume that $\langle R, X \rangle$, $\langle R_1, X_1 \rangle$ generate the same consequence operation and let $r \in \mathrm{Der}(R, X)$. Thus, by Lemma 1.47 (ii),

$$\alpha \in Cn(R, X \cup \Pi) = Cn(R_1, X_1 \cup \Pi) \quad \text{for all} \quad \langle \Pi, \alpha \rangle \in r.$$

We conclude that $r \in \mathrm{Der}(R_1, X_1)$ and hence $\mathrm{Der}(R, X) \subseteq \mathrm{Der}(R_1, X_1)$. The reverse inclusion can be shown by a similar argument.

Let us assume, on the other hand, that $\mathrm{Der}(R, X) = \mathrm{Der}(R_1, X_1)$ and suppose $\alpha \in Cn(R, X \cup Y)$ for some $\alpha \in S$, $Y \subseteq S$ (some finite $Y \subseteq S$).

Let $\beta \in X_1$ and consider the one-element rule $r = \{\langle Y \cup \{\beta\}, \alpha \rangle\}$. Since $\alpha \in Cn(R, X \cup Y) \subseteq Cn(R, X \cup Y \cup \{\beta\})$, we infer that the rule r is derivable in $\langle R, X \rangle$ — see Lemma 1.47 (ii). Hence $r \in \mathrm{Der}(R_1, X_1)$, which yields $\alpha \in Cn(R_1, X_1 \cup Y \cup \{\beta\}) = Cn(R_1, X_1 \cup Y)$. Thus, $Cn(R, X \cup Y) \subseteq Cn(R_1, X_1 \cup Y)$ for each $Y \subseteq S$ (each finite $Y \subseteq S$). The proof of the reverse inclusion is analogous. $\square$

The assumption that X, X_1 are non-empty is necessary for the implications ($\Rightarrow$). This is a result of the exclusion of axiomatic rules — see commentary after Definition 1.1. The signs of equality can be replaced in Lemma 1.50 (ii) by the inclusion. Such operation will be, however, forbidden in Lemma 1.50 (i). The inclusion $\mathrm{Adm}(R, X) \subseteq \mathrm{Adm}(R_1, X_1)$ yields neither $Cn(R, X) \subseteq Cn(R_1, X_1)$ nor $Cn(R_1, X_1) \subseteq Cn(R, X)$. The inclusion $Cn(R, X) \subseteq Cn(R_1, X_1)$ does not yield any inclusion between $\mathrm{Adm}(R, X)$ and $\mathrm{Adm}(R_1, X_1)$.

The pair $\langle R, X \rangle$ is a *subsystem* of $\langle R_1, X_1 \rangle$, in symbols $\langle R, X \rangle \preccurlyeq \langle R_1, X_1 \rangle$, if and only if $X \subseteq Cn(R_1, X_1)$ and $R \subseteq \mathrm{Der}(R_1, X_1)$.

If $\langle R, X \rangle$ is a subsystem of $\langle R_1, X_1 \rangle$, then we obtain

(i) $\mathrm{Der}(R, X) \subseteq \mathrm{Der}(R_1, X_1)$,

(ii) $Cn(R, X) \subseteq Cn(R_1, X_1)$;

i.e., all formulas and rules which are derivable from R and X are also derivable in the system $\langle R_1, X_1 \rangle$. It should be noticed that $\mathrm{Adm}(R, X) \subseteq \mathrm{Adm}(R_1, X_1)$ is not brought about by the relation $\preccurlyeq$ (nor is the reverse inclusion).

If we assume additionally that X, X_1 are non-empty, then the condition (i) implies (ii) and hence

$$\langle R, X \rangle \preccurlyeq \langle R_1, X_1 \rangle \quad \Leftrightarrow \quad \mathrm{Der}(R, X) \subseteq \mathrm{Der}(R_1, X_1).$$

It is also clear that the relation $\preccurlyeq$ is reflexive and transitive and hence

$$\langle R, X \rangle \approx \langle R_1, X_1 \rangle \quad \Leftrightarrow \quad \langle R, X \rangle \preccurlyeq \langle R_1, X_1 \rangle \ \wedge \ \langle R_1, X_1 \rangle \preccurlyeq \langle R, X \rangle$$

is an equivalence relation. The equivalence of $\langle R, X \rangle$ and $\langle R_1, X_1 \rangle$ yields:

(i) $\mathrm{Der}(R, X) = \mathrm{Der}(R_1, X_1)$;

(ii) $Cn(R, X) = Cn(R_1, X_1)$;

(iii) $\mathrm{Adm}(R, X) = \mathrm{Adm}(R_1, X_1)$.

Assuming that X, X_1 are non-empty, one can prove

$$\langle R, X \rangle \approx \langle R_1, X_1 \rangle \quad \Leftrightarrow \quad \mathrm{Der}(R, X) = \mathrm{Der}(R_1, X_1).$$

Using some properties of Adm and Der, we can state now

$$\langle R, X \rangle \approx \langle \mathrm{Der}(R, X), X \rangle \approx \langle R, Cn(R, X) \rangle \approx \langle \mathrm{Der}(R, X), Cn(R, X) \rangle$$
$$\preccurlyeq \langle \mathrm{Adm}(R, X), X \rangle \approx \langle \mathrm{Adm}(R, X), Cn(R, X) \rangle.$$

We will write $\langle R, X \rangle \prec \langle R_1, X_1 \rangle$ when $\langle R, X \rangle \preccurlyeq \langle R_1, X_1 \rangle$ and $\langle R, X \rangle \not\approx \langle R_1, X_1 \rangle$. It should be noted that in most cases

$$\langle R, X \rangle \prec \langle \mathrm{Adm}(R, X), Cn(R, X) \rangle.$$

The relations $\preccurlyeq$, $\approx$ can also be defined in terms of consequence operations.

Lemma 1.51. *For every $X, X_1 \subseteq S$ and every $R, R_1 \subseteq \mathscr{R}_S$:*

(i) $\langle R, X \rangle \preccurlyeq \langle R_1, X_1 \rangle \quad \Leftrightarrow \quad Cn_{RX} \leqslant Cn_{R_1 X_1}$;

(ii) $\langle R, X \rangle \approx \langle R_1, X_1 \rangle \quad \Leftrightarrow \quad Cn_{RX} = Cn_{R_1 X_1}$.

One can say that the lattice ordering $\leqslant$ on the family of all consequence operations is induced by the relation $\preccurlyeq$. Similarly, other definitions as introduced for pairs $\langle R, X \rangle$ can also be formulated for consequence operations, e.g., the rule r is said to be derivable (admissible) with respect to a consequence Cn, in symbols $r \in DER(Cn)$ $(r \in ADM(Cn))$ iff $\alpha \in Cn(\Pi)$ (iff $\Pi \subseteq Cn(\emptyset) \Rightarrow \alpha \in Cn(\emptyset)$) for every $\langle \Pi, \alpha \rangle \in r$.

From the family of all systems $\langle R, X \rangle$ generating a given consequence operation Cn we choose the pair $\langle DER(Cn), Cn(\emptyset) \rangle$, which will be called a *closed system* of a propositional logic. Any such system can also be considered in the form $\langle \mathrm{Der}(R, X), Cn(R, X) \rangle$ for some $R \subseteq \mathscr{R}_S$ and $X \subseteq S$. Obviously,

$$\langle R, X \rangle \approx \langle \mathrm{Der}(R, X), Cn(R, X) \rangle.$$

Given an indexed family $\{\langle R_t, X_t \rangle : t \in T\}$ of propositional systems we define

$$\prod_{t \in T} \langle R_t, X_t \rangle = \langle \bigcap_{t \in T} \mathrm{Der}(R_t, X_t), \bigcap_{t \in T} Cn(R_t, X_t) \rangle,$$

$$\coprod_{t \in T} \langle R_t, X_t \rangle = \langle \bigcup_{t \in T} R_t, \bigcup_{t \in T} X_t \rangle.$$

Note that $\prod_{t \in T} \langle R_t, X_t \rangle$ is a closed system, i.e.,

Lemma 1.52.

(i) $\mathrm{Der}\big(\prod_{t \in T} \langle R_t, X_t \rangle \big) = \bigcap_{t \in T} \mathrm{Der}(R_t, X_t)$;

(ii) $Cn\big(\prod_{t \in T} \langle R_t, X_t \rangle \big) = \bigcap_{t \in T} Cn(R_t, X_t)$.

The operations $\coprod$, $\prod$ correspond to the lattice-theoretical operations of supremum and infimum in the family of all consequences (finitistic consequences).

Structurality

Any $e : At \to S$ can be uniquely extended to an endomorphism $h^e : S \to S$, see Lemma 1.4. The mapping e is usually called a *substitution*. Let $\alpha = \alpha(p_1, \ldots, p_k)$, that is $At(\alpha) \subseteq \{p_1,,\ldots,p_k\}$, and $e(p_i) = \gamma_i$ for each i. The result of the substitution in the formula α (i.e., the formula $h^e(\alpha)$) depends in fact on the values the substitution takes on the variables occurring in α. For this reason, the result of the substitution is sometimes noted in the form $\alpha[p_1/\gamma_1 \ldots p_k/\gamma_k]$ or even as $\alpha(\gamma_1, \ldots, \gamma_k)$. We have

$$h^e(\alpha) = \alpha[p_1/\gamma_1 \ldots p_k/\gamma_k] = \alpha(\gamma_1, \ldots, \gamma_k).$$

The *substitution rule* over S is defined by

$$\langle \{\alpha\}, \beta \rangle \in r_* \quad \Leftrightarrow \quad (\beta = h^e(\alpha), \text{ for some } e : At \to S)$$

or by the scheme

$$r_* : \frac{\alpha}{h^e(\alpha)} \quad \text{for all } \alpha \in S \text{ and all } e : At \to S$$

and the consequence operation based on the substitution rule is denoted by Sb. We have, for each $X \subseteq S$,

$$Sb(X) = Cn(\{r_*\}, X) = \{\alpha : \alpha \in h^e(X), \text{ for some } e : At \to S\}.$$

A rule $r \in \mathscr{R}_S$ is said to be *structural*, in symbols $r \in \mathrm{Struct}$, iff

$$\langle \Pi, \alpha \rangle \in r \quad \Rightarrow \quad \langle h^e(\Pi), h^e(\alpha) \rangle \in r, \quad \text{for all } e : At \to S.$$

The definition of structurality is due to J. Łoś and R. Suszko [64], 1958. One could say that any reasoning based on structural rules preserves its validity in any similar situation (which means, is preserved under substitutions). All propositional rules occurring in formal logic (e.g., modus ponens) are structural — the only exception is the substitution rule!

A system $\langle R, X \rangle$, where $R \subseteq \mathscr{R}_S$ and $X \subseteq S$, is called *invariant*, in symbols $\langle R, X \rangle \in \text{Inv}$, if $R \subseteq \text{Struct}$ and $X = Sb(X)$. If $R \subseteq \text{Struct}$, then $\langle R \cup \{r_*\}, X \rangle$ is said to be *substitutional*.

A sequent $\langle \Pi, \alpha \rangle$ is called a *basic sequent* of a structural rule r' iff

$$r' = \{ \langle h^e(\Pi), h^e(\alpha) \rangle : \text{ for all substitutions } e \}.$$

If r' has a basic sequent, then it is said to be a *standard* rule. Let us observe that $\langle \{p \to q, p\}, q \rangle$ is a basic sequent of the modus ponens rule. One could always replace any structural rule, say r, with its standard subrules as

$$r = \bigcup \{r' : r' \subseteq r \text{ and } r' \text{ is standard}\}.$$

Obviously, any standard rule is structural but not conversely. As an important example of a structural rule without basic sequent, let us introduce the so-called *'big rule'*, denoted by r_X, for each $X \subseteq S$.

Definition 1.53. For each $X, \Pi \subseteq S$ and each $\alpha \in S$,

$$\langle \Pi, \alpha \rangle \in r_X \Leftrightarrow \left(h^e(\Pi) \subseteq X \Rightarrow h^e(\alpha) \in X, \text{ for all substitutions } e \right).$$

It is easily seen that r_X is structural but it need not be standard. For example, $\langle \{p + q\}, p \rangle \in r_{\{p\}}$, $\langle \{p \to q\}, p \rangle \in r_{\{p\}}$ but $\langle \{q\}, p \rangle \notin r_{\{p\}}$.

The concept of structurality (and that of invariantness) is in some sense too restricted. For example, the rule

$$r : \frac{\alpha(p)}{\alpha(q)} \qquad \text{where } \alpha \quad \text{contains one variable and } p, q \in At$$

is not structural. It is, however, preserved under all substitutions $e : At \to S$ such that $e(At) \subseteq At$. It proves to be useful to have a concept of structurality restricted to a certain set $\Gamma \subseteq S$. Thus, let us define

$$r_* | \Gamma : \frac{\alpha}{h^e(\alpha)} \qquad \text{for all } \alpha \in S \text{ and all } e : At \to \Gamma.$$

Then we take $Sb_\Gamma(X) = Cn(\{r_* | \Gamma\}, X)$. Of course, $r_* | S = r_*$, $Sb_S = Sb$ and $r_* | \emptyset = \emptyset$, $Sb_\emptyset(X) = X$ for every $X \subseteq S$. It is also easy to see that

$$Sb_\Gamma(X \cup Y) = Sb_\Gamma(X) \cup Sb_\Gamma(Y), \quad \text{for each } X, Y \subseteq S.$$

Definition 1.54. A rule r is said to be Γ-*structural*, in symbols $r \in \text{Struct}(\Gamma)$, if

$$\langle \Pi, \alpha \rangle \in r \Rightarrow \langle h^e(\Pi), h^e(\alpha) \rangle \in r, \quad \text{for all} \ \ e : At \to \Gamma.$$

We say that a system $\langle R, X \rangle$ is Γ-*invariant*, in symbols $\langle R, X \rangle \in \Gamma - \text{Inv}$, iff $R \subseteq \text{Struct}(\Gamma)$ and $X = Sb_\Gamma(X)$. Obviously, the concept of S-structurality (S-invariantness) coincides with structurality (invariantness), i.e.,

$$\text{Struct} = \text{Struct}(S) \qquad \text{and} \qquad \text{Inv} = S - \text{Inv}.$$

On the other hand, all rules over S are $\emptyset$-structural and hence $\langle R, X \rangle \in \emptyset - \text{Inv}$ for every $R \subseteq \mathscr{R}_S$ and $X \subseteq S$.

Lemma 1.55. *For every* $\Gamma_1, \Gamma_2 \subseteq S$:

(i) $\Gamma_1 \subseteq \Gamma_2 \quad \Rightarrow \quad \text{Struct}(\Gamma_2) \subseteq \text{Struct}(\Gamma_1)$;

(ii) $\Gamma_1 \subseteq \Gamma_2 \quad \Rightarrow \quad \Gamma_2 - \text{Inv} \subseteq \Gamma_1 - \text{Inv}$.

The proof is very easy. It follows from Lemma 1.55 that for every $\Gamma \subseteq S$,

$$\text{Struct} \subseteq \text{Struct}(\Gamma) \subseteq \mathscr{R}_S \qquad \text{and} \qquad \text{Inv} \subseteq \Gamma - \text{Inv} \subseteq \emptyset - \text{Inv}.$$

Lemma 1.56. *If* $R \subseteq \text{Struct}(\Gamma)$ *and* $X \subseteq S$, *then*

$$h^e\big(Cn(R, X)\big) \subseteq Cn\big(R, h^e(X)\big), \quad \text{for every} \ \ e : At \to \Gamma.$$

We have left this lemma without proof.

Corollary 1.57. *If* $R \subseteq \text{Struct}(\Gamma)$ *and* $X, Y \subseteq S$, *then*

$$h^e\big(Cn(R, Sb_\Gamma(X) \cup Y)\big) \subseteq Cn\big(R, Sb_\Gamma(X) \cup h^e(Y)\big), \quad \text{for every} \ \ e : At \to \Gamma.$$

With the help of Lemma 1.56, we obtain also the following theorem on the reduction of the substitution rule to the set of axioms:

Theorem 1.58. *If* $R \subseteq \text{Struct}(\Gamma)$ *and* $X \subseteq S$, *then*

$$Cn\big(R \cup \{r_*|\Gamma\}, X\big) = Cn\big(R, Sb_\Gamma(X)\big).$$

Proof. We have, of course, $Cn\big(R, Sb_\Gamma(X)\big) \subseteq Cn\big(R \cup \{r_*|\Gamma\}, X\big)$. To prove the reverse inclusion let $\alpha \in Cn\big(R, Sb_\Gamma(X)\big)$. By Lemma 1.56, $h^e(\alpha) \in Cn\big(R, h^e(Sb_\Gamma(X))\big)$ if $e : At \to \Gamma$. Since $h^e(Sb_\Gamma(X)) \subseteq Sb_\Gamma(X)$, we conclude that $\beta \in Cn\big(R, Sb_\Gamma(X)\big)$ for each $\beta \in S$ such that $\langle \{\alpha\}, \beta \rangle \in r_*|\Gamma$. It shows that the set $Cn\big(R, Sb_\Gamma(X)\big)$ is closed under the rule $r_*|\Gamma$. Since the same set is closed under the rules R, then $Cn\big(R \cup \{r_*|\Gamma\}, X\big) \subseteq Cn\big(R, Sb_\Gamma(X)\big)$ on the basis of Definition 1.40. $\square$

From the above it follows that, for every $X \subseteq S$ and $R \subseteq \text{Struct}(\Gamma)$,

$$Sb_\Gamma\big(Cn(R, X)\big) \subseteq Cn\big(R, Sb_\Gamma(X)\big).$$

Thus, each Γ-invariant system $\langle R, X \rangle$ is closed under the rule $r_*|\Gamma$, which also means that $r_*|\Gamma \in \text{Adm}(R, Sb_\Gamma(X))$.

Corollary 1.59. *If* $X \subseteq S, r \in \mathscr{R}_S$ *and* $R \subseteq \mathrm{Struct}(\Gamma)$, *then*

$$r \in \mathrm{Der}(R \cup \{r_*|\Gamma\}, X) \Leftrightarrow \big(r \in \mathrm{Adm}(R, Sb_\Gamma(X) \cup Sb_\Gamma(Y)\big), \quad \text{for all } Y \subseteq S\big).$$

Let us refer to our remark on non-structurality of the substitution rule. It appears that r_* is not only non-structural but also non-derivable in any consistent system $\langle R, Sb(X)\rangle$ with structural rules R and with non-empty X. However, the rule r_* will be admissible in any such system.

Theorem 1.60. *For every* $R \subseteq \mathrm{Struct}$ *and every* $X \subseteq S$:

(i) $r_* \in \mathrm{Adm}(R, Sb(X))$;

(ii) $r_* \notin \mathrm{Der}(R, Sb(X))$ *if* $\emptyset \neq Cn\big(R, Sb(X)\big) \neq S$.

Proof. By Corollary 1.59 it suffices to show (ii) only. Let us notice that $\langle\{p\}, q\rangle \in r_*$ for every $p, q \in At$. If $e : At \to S$ is a substitution such that $e(p) \in Cn\big(R, Sb(X)\big)$ and $e(q) \notin Cn\big(R, Sb(X)\big)$, then $h^e(q) \notin Cn\big(R, Sb(X) \cup \{h^e(p)\}\big)$ and hence, by Lemma 1.56, $q \notin Cn\big(R, Sb(X) \cup \{p\}\big)$. Then, according to Lemma 1.47 (ii), $r_* \notin \mathrm{Der}(R, Sb(X))$. $\qquad\square$

The addition of the substitution rule to an invariant system $\langle R, X\rangle$ does not change the set of formulas derivable in $\langle R, X\rangle$. The systems $\langle R, X\rangle$, $\langle R \cup \{r_*\}, X\rangle$ will be, however, non-equivalent. Some propositional logics, as for example the classical logic, will be further considered in two non-equivalent versions: invariant $\langle R, Sb(X)\rangle$, where $R \subseteq \mathrm{Struct}$, and substitutional $\langle R \cup \{r_*\}, X\rangle$.

Let us proceed to the consequence formalism of propositional logics. Instead of Γ-structurality of rules and Γ-invariantness of propositional systems we can speak about Γ-structurality of consequence operations.

A consequence Cn is said to be Γ-structural, in symbols $Cn \in \mathrm{STRUCT}(\Gamma)$, provided that $h^e(Cn(X)) \subseteq Cn(h^e(X))$ for each $e : At \to \Gamma$ and each $X \subseteq S$.

Similarly, Cn is called structural, $Cn \in \mathrm{STRUCT}$ (cf. [64], 1958) if and only if $h^e(Cn(X)) \subseteq Cn(h^e(X))$ for each $X \subseteq S$ and each $e : At \to S$.

On the basis of Corollary 1.57 any Γ-invariant system $\langle R, X\rangle$ determines a Γ-structural consequence Cn_{RX}. On the other hand, any Γ-structural consequence is generated by some Γ-invariant system (see the proof of Theorem 1.43).

Without any proof let us note that the family of all structural consequences (Γ-structural consequences) forms a complete sublattice of the lattice of all consequences over S, see [140], 1970, i.e.,

Lemma 1.61. *If* $\{Cn_t : t \in T\}$ *is an indexed family of* Γ-*structural (structural) consequences, then the operations* $\prod_{t \in T} Cn_t$ *and* $\coprod_{t \in T} Cn_t$ *are also* Γ-*structural (structural).*

1.5 Brief exposition of the most important propositional logics

To make some of the notions introduced in this chapter more familiar, we discuss in this section concrete examples of propositional logics.

Intuitionistic logic

This logic is formalized in the standard language $\mathscr{S}_2 = \langle S_2, \rightarrow, +, \cdot, \sim \rangle$. Let A_i be the set of the following axioms,

(1) $p \rightarrow (q \rightarrow p)$

(2) $(p \rightarrow (p \rightarrow q)) \rightarrow (p \rightarrow q)$

(3) $(p \rightarrow q) \rightarrow [(q \rightarrow s) \rightarrow (p \rightarrow s)]$

(4) $p \rightarrow p + q$

(5) $q \rightarrow p + q$

(6) $(p \rightarrow s) \rightarrow ((q \rightarrow s) \rightarrow (p + q \rightarrow s))$

(7) $p \cdot q \rightarrow p$

(8) $p \cdot q \rightarrow q$

(9) $(p \rightarrow q) \rightarrow [(p \rightarrow r) \rightarrow (p \rightarrow q \cdot r)]$

(10) $p \rightarrow (\sim p \rightarrow q)$

(11) $(p \rightarrow \sim p) \rightarrow \sim p$

and let $R_{0*} = \{r_0, r_*\}$, where r_0 is the modus ponens and r_* is the substitution rule over S_2. Then $\langle R_{0*}, A_i \rangle$ is a system of *intuitionistic propositional logic*. This logic will also be considered in the invariant version $\langle R_0, Sb(A_i) \rangle$, where $R_0 = \{r_0\}$ and $Sb(A_i)$ contains all substitutions of the formulas from A_i. By Theorem 1.58,

$$Cn(R_{0*}, A_i) = Cn(R_0, Sb(A_i)).$$

Let Cn_i be the consequence operation generated by $\langle R_0, Sb(A_i) \rangle$, i.e.,

$$Cn_i(X) = Cn(R_0, Sb(A_i) \cup X) \quad \text{for each} \ \ X \subseteq S_2.$$

One of the most important properties of intuitionistic logic (and many other systems) is the deduction theorem:

Theorem 1.62. *For every $X \subseteq S_2$ and every $\alpha, \beta \in S_2$,*

$$\beta \in Cn_i(X \cup \{\alpha\}) \quad \Leftrightarrow \quad (\alpha \rightarrow \beta) \in Cn_i(X).$$

The deduction theorem can also be viewed as a characterization of the intuitionistic implication. The other connectives of intuitionistic logic can be characterized in a similar way.

Theorem 1.63. *For every $X \subseteq S_2$ and every $\alpha, \beta \in S_2$:*

(i) $Cn_i(X \cup \{\alpha \cdot \beta\}) = Cn_i(X \cup \{\alpha, \beta\})$;

(ii) $Cn_i(X \cup \{\alpha + \beta\}) = Cn_i(X \cup \{\alpha\}) \cap Cn_i(X \cup \{\beta\})$.

The easy proof based on the axioms (4)–(9) and on the deduction theorem is left to the reader. Observe that the inclusion ($\subseteq$) in (i) states derivability, on the ground of $\langle R_0, Sb(A_i) \rangle$, of the *adjunction rule*

$$r_a : \frac{\alpha, \beta}{\alpha \cdot \beta} \qquad \text{for all} \ \ \alpha, \beta \in S_2.$$

We recall that $\alpha \equiv \beta$ is the abbreviation for $(\alpha \to \beta) \cdot (\beta \to \alpha)$. From Theorems 1.62 and 1.63 we derive

Corollary 1.64. *For every $X \subseteq S_2$ and every $\alpha, \beta \in S_2$,*

$$(\alpha \equiv \beta) \in Cn_i(X) \quad \Leftrightarrow \quad Cn_i(X \cup \{\alpha\}) = Cn_i(X \cup \{\beta\}).$$

The characterization of the intuitionistic negation is as follows.

Theorem 1.65. *For every $X \subseteq S_2$ and every $\alpha \in S_2$,*

$$\sim \alpha \in Cn_i(X) \quad \Leftrightarrow \quad Cn_i(X \cup \{\alpha\}) = S_2.$$

On the grounds of 1.62–1.65, we get

Corollary 1.66. *For every $X \subseteq S_2$ and every $\alpha, \beta \in S_2$:*

(i) $\alpha \to \beta \in Cn_i(X) \quad \Leftrightarrow \quad Cn_i(X \cup \{\beta\}) \subseteq Cn_i(X \cup \{\alpha\})$;

(ii) $\alpha + \beta \in Cn_i(X) \quad \Leftrightarrow \quad Cn_i(X \cup \{\alpha\}) \cap Cn_i(X \cup \{\beta\}) \subseteq Cn_i(X)$;

(iii) $\alpha \cdot \beta \in Cn_i(X) \quad \Leftrightarrow \quad Cn_i(X \cup \{\alpha\}) \cup Cn_i(X \cup \{\beta\}) \subseteq Cn_i(X)$;

(iv) $\sim \alpha \in Cn_i(X) \quad \Leftrightarrow \quad S_2 \subseteq Cn_i(X \cup \{\alpha\})$.

One can show (see [87], 1960) that Cn_i is the least consequence operation over S_2 satisfying the above conditions (i)–(iv), i.e., Cn_i fulfills conditions (i)–(iv) and if a consequence operation Cn fulfills (i)–(iv) then $Cn_i \leqslant Cn$.

Assume that S is an implicational sublanguage of S_2. Wajsberg's separation theorem says that

$$S \cap Cn_i(\emptyset) = Cn(R_0, S \cap Sb(A_i)).$$

We pay special attention to the positive fragment of the intuitionistic logic, i.e., to the *logic of Hilbert*. The set of axioms of Hilbert's logic will be further denoted by A_H, that is $A_H = S_1 \cap A_i$. We will also consider the pure implicational fragment of intuitionistic logic, that is the pure implicational logic of Hilbert. This logic is determined by the modus ponens and the substitution rule formalized in the language $\mathscr{S}^{\to} = \langle S^{\to}, \to \rangle$ and by the formulas $A_H^{\to} = A_i \cap S^{\to} = A_H \cap S^{\to}$.

Classical logic

Let $A_2 = A_i \cup \{\sim\sim p \to p\}$ where A_i is the set of axioms for intuitionistic propositional logic. The system $\langle R_0, Sb(A_2) \rangle$ (or $\langle R_{0*}, A_2 \rangle$) is called the *classical propositional logic*. Obviously

$$Cn(R_{0*}, A_i) \subseteq Cn(R_{0*}, A_2).$$

Let Cn_2 be the consequence operation generated by $\langle R_0, Sb(A_2) \rangle$, i.e.,

$$Cn_2(X) = Cn(R_0, Sb(A_2) \cup X), \quad \text{for each } X \subseteq S_2.$$

The system of classical logic will play a privileged role in our considerations. This system enjoys many specific properties. In particular, it follows from Corollary 1.66 that

Corollary 1.67. *For each $X \subseteq S_2$ and each $\alpha, \beta \in S_2$:*

(i) $(\alpha \to \beta) \in Cn_2(X) \quad \Leftrightarrow \quad \beta \in Cn_2(X \cup \{\alpha\})$;

(ii) $Cn_2(X \cup \{\alpha \cdot \beta\}) = Cn_2(X \cup \{\alpha, \beta\})$;

(iii) $Cn_2(X \cup \{\alpha + \beta\}) = Cn_2(X \cup \{\alpha\}) \cap Cn_2(X \cup \{\beta\})$;

(iv) $\sim \alpha \in Cn_2(X) \quad \Leftrightarrow \quad Cn_2(X \cup \{\alpha\}) = S_2$;

(v) $\alpha \in Cn_2(X) \quad \Leftrightarrow \quad Cn_2(X \cup \{\sim \alpha\}) = S_2$.

One can show (see [77], 1969) that the operation determined by $\langle R_0, Sb(A_2) \rangle$ is the greatest structural and consistent consequence operation satisfying the above conditions (i)–(iv). Since, on the other hand, $\langle R_0, Sb(A_2) \rangle$ is the least system satisfying the conditions (i)–(v), Corollary 1.67 determines uniquely the classical propositional logic: a system $\langle R, A \rangle \in \text{Inv} \cap \text{Cns}$ fulfills (i)–(v) if and only if $\langle R, A \rangle \approx \langle R_0, Sb(A_2) \rangle$ Conditions (i)–(v) have been formulated for the classical consequence operation by A. Tarski [117], 1930.

It should be remembered that the axioms A_2 of the classical logic that we have presented are not separable, since

$$S \cap Cn(R_0, Sb(A_2)) \neq Cn(R_0, Sb(A_2) \cap S)$$

for any proper implicational sublanguage S of S_2. The formula $((p \to q) \to p) \to p$, for example, is not derivable in Hilbert's logic and hence it cannot be deduced by means of r_0 from the positive axioms of $\langle R_0, Sb(A_2) \rangle$.

We can axiomatize all implicative fragments of classical logic by adjoining to the axioms $A_2 \cap S$ the formula $((p \to q) \to p) \to p$. Sets of axioms that are thus obtained will be denoted by $A_2^{\to *}$, $A_2^{\to}$, $A_2^{\to +}$ respectively.

The modal system $S5$

The *modal system* $S5$, can be defined over S_2 by the following set of axioms,

A_{S5}:   (1) $(p \to (q \to s)) \to ((p \to q) \to (p \to s))$

(2) $(s \to t) \to (q \to (s \to t))$

(3) $p \cdot q \to p$

(4) $p \cdot q \to q$

(5) $(p \to q) \to ((p \to s) \to (p \to q \cdot s))$

(6) $p \to p + q$

(7) $q \to p + q$

(8) $(p \to s) \to ((q \to s) \to (p + q \to s))$

(9) $\sim (s \to t) \to ((s \to t) \to q)$

(10) $p \cdot \sim p \to q$

(11) $(p \to \sim p) \to \sim p$

(12) $p \cdot \sim (p \cdot q) \to \sim q$

(13) $(p \to \sim\sim p) \cdot (\sim\sim p \to p)$

and rules: r_0 (modus ponens), r_* (substitution), r_a (adjunction). Let us take $R_{0a*} = \{r_0, r_a, r_*\}$ and $R_{0a} = \{r_0, r_a\}$. Obviously, according to Theorem 1.58,

$$Cn\big(R_{0a*}, A_{S5}\big) = Cn\big(R_{0a}, Sb(A_{S5})\big).$$

We shall now write down the deduction theorem for $S5$.

Theorem 1.68. *If $X \subseteq \{\varphi \to \psi : \varphi, \psi \in S_2\}$, then for every $\alpha, \beta \in S_2$,*

$$\beta \in Cn\big(R_{0a}, Sb(A_{S5}) \cup X \cup \{\alpha\}\big) \; \Rightarrow \; (\alpha \to \beta) \in Cn\big(R_{0a}, Sb(A_{S5}) \cup X\big).$$

Proof. Let $\alpha \in S_2$ and consider the set

$$Y = \{\beta \in S_2 : \alpha \to \beta \in Cn\big(R_{0a}, Sb(A_{S5}) \cup X\big)\}.$$

First, we prove that $Sb(A_{S5}) \cup X \cup \{\alpha\} \subseteq Y$. If $\beta \in Sb(A_{S5}) \cup X$, then by (2) and r_0 we obtain $\alpha \to \beta \in Cn\big(R_0, Sb(A_{S5}) \cup X\big)$, hence $\beta \in Y$. If $\alpha = \beta$, then

$$\alpha \to \beta \in Sb(p \to p) \subseteq Cn\big(R_{0a}, Sb(A_{S5})\big) \subseteq Cn\big(R_{0a}, Sb(A_{S5}) \cup X\big).$$

Thus, $Sb(A_{S5}) \cup X \cup \{\alpha\} \subseteq Y$.

Now, let us try to show that the set Y is closed under the rule r_0, i.e.,

$$\beta \, , \, \beta \to \gamma \in Y \qquad \Rightarrow \qquad \gamma \in Y.$$

Assume that $\alpha \to \beta$ and $\alpha \to (\beta \to \gamma)$ belong to $Cn(R_{0a}, Sb(A_{S5}) \cup X)$. Since $(\alpha \to (\beta \to \gamma)) \to ((\alpha \to \beta) \to (\alpha \to \gamma))$ is a substitution of the first axiom and since $Cn(R_{0a}, Sb(A_{S5}) \cup X)$ is closed under r_0, it follows that $\alpha \to \gamma \in Cn(R_{0a}, Sb(A_{S5}) \cup X)$. Thus, $\gamma \in Y$ which was to be proved.

Similarly, it can be shown that Y is closed under r_a since, by (5),

$$\alpha \to \beta, \alpha \to \gamma \in Cn(R_{0a}, Sb(A_{S5}) \cup X) \ \Rightarrow \ (\alpha \to \beta \cdot \gamma) \in Cn(R_{0a}, Sb(A_{S5}) \cup X).$$

Then, in the light of Definition 1.40, we have

$$Cn(R_{0a}, Sb(A_{S5}) \cup X \cup \{\alpha\}) \subseteq Y,$$

i.e., if $\beta \in Cn(R_{0a}, Sb(A_{S5}) \cup X \cup \{\alpha\})$, then $\alpha \to \beta \in Cn(R_{0a}, Sb(A_{S5}) \cup X)$. $\square$

Observe that the reverse implication holds for every $X \subseteq S_2$. From Theorem 1.68 it immediately follows that the following formulas are derivable in $S5$.

(14) $(p \to q) \to ((q \to s) \to (p \to s))$

(15) $(q \to s) \to ((p \to q) \to (p \to s))$

(16) $((p \to p) \to p) \to p$

(17) $q \to (p \to p)$

Let $\beta_1 \cdot \ldots \cdot \beta_k$ denote the formula $(\ldots (\beta_1 \cdot \beta_2) \cdot \ldots) \cdot \beta_k$. The deduction theorem for $S5$ can also be formulated as follows.

Corollary 1.69. *For every $X \subseteq S_2$ and every $\alpha \in S_2$,*

$\alpha \in Cn(R_{0a}, Sb(A_{S5}) \cup X)$

$\Leftrightarrow (\beta_1 \cdot \ldots \cdot \beta_k \to \alpha) \in Cn(R_{0a}, Sb(A_{S5}))$, *for some* $\beta_1, \ldots, \beta_k \in X$.

The characterization of negation in $S5$ is similar to that in classical logic.

Theorem 1.70. *For every $X \subseteq S_2$ and every $\alpha \in S_2$:*

(i) $\sim \alpha \in Cn(R_{0a}, Sb(A_{S5}) \cup X) \Leftrightarrow Cn(R_{0a}, Sb(A_{S5}) \cup X \cup \{\alpha\}) = S_2$;

(ii) $\alpha \in Cn(R_{0a}, Sb(A_{S5}) \cup X) \Leftrightarrow Cn(R_{0a}, Sb(A_{S5}) \cup X \cup \{\sim \alpha\}) = S_2$.

Proof. We will prove (i) only since (ii) follows from (i) and axiom (13). Moreover, let us observe that the implication ($\Rightarrow$) in (i) is a simple consequence of the adjunction rule and (10).

Assume that $Cn(R_{0a}, Sb(A_{S5}) \cup X \cup \{\alpha\}) = S_2$. Then

$$p \cdot \sim p \in Cn(R_{0a}, Sb(A_{S5}) \cup \{\beta_1, \ldots, \beta_k\} \cup \{\alpha\})$$

for some $\beta_1, \ldots, \beta_k \in X$. Hence, by (10),

$$Cn(R_{0a}, Sb(A_{S5}) \cup \{\beta_1, \ldots, \beta_k, \alpha\}) = S_2.$$

Using the deduction theorem Corollary 1.69 we obtain then

$$(\beta_1 \cdot \ldots \cdot \beta_k \cdot \alpha) \to\sim (\beta_1 \cdot \ldots \cdot \beta_k \cdot \alpha) \in Cn\big(R_{0a}, Sb(A_{S5})\big)$$

and hence, by (11) and r_a,

$$(\beta_1 \cdot \ldots \cdot \beta_k) \cdot \sim (\beta_1 \cdot \ldots \cdot \beta_k \cdot \alpha) \in Cn\big(R_{0a}, Sb(A_{S5}) \cup X\big).$$

Then it follows from (12) that $\sim \alpha \in Cn\big(R_{0a}, Sb(A_{S5}) \cup X\big)$. $\square$

   Then, using Theorems 1.68 and 1.70, we can derive the following formulas.

(18) $(p \to q) \to (\sim q \to\sim p)$

(19) $(p \to\sim q) \to (q \to\sim p)$

(20) $(\sim p \to q) \to (\sim q \to p)$

(21) $(\sim p \to\sim q) \to (q \to p)$

(22) $\sim (s \to t) \to (q \to\sim (s \to t))$ by (9) and (19) .

Theorem 1.71. *For every $X \subseteq S_2$ and every $\alpha, \beta \in S_2$:*

   (i) $Cn\big(R_{0a}, Sb(A_{S5}) \cup X \cup \{\alpha \cdot \beta\}\big) = Cn\big(R_{0a}, Sb(A_{S5}) \cup X \cup \{\alpha, \beta\}\big);$

   (ii) $Cn\big(R_{0a}, Sb(A_{S5}) \cup X \cup \{\alpha + \beta\}\big) = \;\; Cn\big(R_{0a}, Sb(A_{S5}) \cup X \cup \{\alpha\}\big)$
$$\cap \; Cn\big(R_{0a}, Sb(A_{S5}) \cup X \cup \{\beta\}\big).$$

Proof. (i) and the inclusion $\subseteq$ in (ii) are obvious. If X is empty, then also the inclusion $\supseteq$ in (ii) holds on the basis of Theorem 1.68 and (8).
   Assume that $X \neq \emptyset$ and let

$$\varphi \in Cn\big(R_{0a}, Sb(A_{S5}) \cup X \cup \{\alpha\}\big) \cap Cn\big(R_{0a}, Sb(A_{S5}) \cup X \cup \{\beta\}\big).$$

Then $\varphi \in Cn\big(R_{0a}, Sb(A_{S5}) \cup \{\gamma_1, \ldots, \gamma_k, \alpha\}\big) \cap Cn\big(R_{0a}, Sb(A_{S5}) \cup \{\gamma_1, \ldots, \gamma_k, \beta\}\big)$
for some $\gamma_1, \ldots, \gamma_k \in X$ and hence, by Theorem 1.70 (ii),

$$S_2 = Cn\big(R_{0a}, Sb(A_{S5}) \cup \{\gamma_1, \ldots, \gamma_k, \alpha, \sim \varphi\}\big)$$
$$= Cn\big(R_{0a}, Sb(A_{S5}) \cup \{\gamma_1, \ldots, \gamma_k, \beta, \sim \varphi\}\big).$$

Using Theorem 1.70 (i) we obtain next

$$\sim (\gamma_1 \cdot \ldots \cdot \gamma_k \cdot \sim \varphi) \in Cn\big(R_{0a}, Sb(A_{S5}) \cup \{\alpha\}\big) \cap Cn\big(R_{0a}, Sb(A_{S5}) \cup \{\beta\}\big).$$

By the deduction theorem and by (8), we have then

$$\sim (\gamma_1 \cdot \ldots \cdot \gamma_k \cdot \sim \varphi) \in Cn\big(R_{0a}, Sb(A_{S5}) \cup \{\alpha + \beta\}\big)$$

and hence — see (12) —

$$\sim\sim \varphi \in Cn\big(R_{0a}, Sb(A_{S5}) \cup X \cup \{\alpha + \beta\}\big).$$ $\square$

The above results allow us to derive in $S5$ the following formulas.

(23) $\sim (p + q) \equiv (\sim p \cdot \sim q)$

(24) $\sim (p \cdot q) \equiv (\sim p + \sim q)$

(25) $p \cdot (\sim p + q) \to q$

(26) $p \cdot q \equiv q \cdot p$

(27) $p \cdot (q \cdot s) \equiv (p \cdot q) \cdot s$

(28) $p + q \equiv q + p$

(29) $p + (q + s) \equiv (p + q) + s$

(30) $p \cdot (q + s) \equiv (p \cdot q) + (p \cdot s)$

(31) $p + (q \cdot s) \equiv (p + q) \cdot (p + s)$

(32) $p + \sim p$

(33) $(p \to p) \cdot q \equiv q$

(34) $\sim (p \to p) \cdot q \equiv \sim (p \to p)$

(35) $(p \to (q \equiv s)) \to (p \cdot q \equiv p \cdot s)$.

Next, let us observe that $\langle R_{0a}, Sb(A_{S5}) \rangle$ is a system with equivalence.

(36) $p \equiv p$

(37) $(p \equiv q) \to (q \equiv p)$

(38) $(p \equiv q) \cdot (q \equiv s) \to (p \equiv s)$

(39) $(p \equiv q) \cdot (s \equiv t) \to (p \to s \equiv q \to t)$

(40) $(p \equiv q) \cdot (s \equiv t) \to (p + s \equiv q + t)$

(41) $(p \equiv q) \cdot (s \equiv t) \to (p \cdot s \equiv q \cdot t)$

(42) $(p \equiv q) \to (\sim p \equiv \sim q)$.

The easy proofs of (36)–(42) are left to the reader. From (36)–(42) immediately follows

Lemma 1.72. *If* $e : At \to S_2$, $f : At \to S_2$ *and* $\alpha \in S_2$, *then*

$$h^e(\alpha) \equiv h^f(\alpha) \in Cn\big(R_{0a}, Sb(A_{S5}) \cup \{e(\gamma) \equiv f(\gamma) : \gamma \in At\}\big).$$

The proof is by induction on the length of α.

Let us introduce the abbreviation

$$\Box \alpha = (\alpha \to \alpha) \to \alpha, \quad \text{for } \alpha \in S_2.$$

It can be easily shown that $\Box \alpha \equiv (\beta \to \beta) \to \alpha \in Cn\big(R_{0a*}, A_{S5}\big)$, see (17), for every $\alpha, \beta \in S_2$. Moreover,

(43) $\Box p \to p$ by (16)

(44) $(p \to q) \to \Box(p \to q)$ by (2)

(45) $\sim \Box p \to \Box \sim \Box p$ by (22)

(46) $\Box p + \Box q \to \Box(p + q)$ by 1.68, 1.71 (ii)

(47) $\Box p \cdot \Box q \to \Box(p \cdot q)$ by 1.68, 1.71 (i).

Next we define the '$\Box$-closed' formulas of S_2,

$$\alpha \in S_\Box \quad \text{iff} \quad \alpha \to \Box\alpha \in Cn(R_{0a*}, A_{S5})$$

or equivalently,

$$\alpha \in S_\Box \quad \text{iff} \quad \alpha \equiv \Box\alpha \in Cn(R_{0a*}, A_{S5}).$$

We list the following properties of $\Box$-closed formulas.

Lemma 1.73. *For every $\alpha, \beta \in S_2$:*

(i) $\alpha \to \beta \in S_\Box$;

(ii) $\alpha \in S_\Box \Rightarrow \sim \alpha \in S_\Box$;

(iii) $\alpha, \beta \in S_\Box \Rightarrow \alpha \cdot \beta, \alpha + \beta \in S_\Box$;

(iv) $Cn(R_{0a*}, A_{S5}) \subseteq S_\Box$.

Proof. Properties (i)–o–(iii) follow from (44)–o–(47). To prove (iv) assume that $\alpha \in \in Cn(R_{0a*}, A_{S5})$, then also $\alpha \in Cn(R_{0a}, Sb(A_{S5}) \cup \{p \to p\})$ and hence, by Theorem 1.68, $\Box\alpha \in Cn(R_{0a}, Sb(A_{S5}))$. $\Box$

As an immediate consequence of Theorem 1.68 we obtain

Corollary 1.74. *For every $X \subseteq S_\Box$ and every $\alpha, \beta \in S_2$,*

$$\beta \in Cn(R_{0a}, Sb(A_{S5}) \cup X \cup \{\alpha\}) \Leftrightarrow (\alpha \to \beta) \in Cn(R_{0a}, Sb(A_{S5}) \cup X).$$

Let us note that $S_\Box$ is a proper subset of S_2 since $\gamma \notin S_\Box$ for any atomic formula γ. Hence, the rule

$$r_\Box : \frac{\alpha}{\Box\alpha} \quad \text{for} \ \alpha \in S_2$$

is not derivable in $\langle R_{0a}, Sb(A_{S5})\rangle$. In the light of Lemma 1.73 (iv) this rule is, however, admissible in $\langle R_{0a}, Sb(A_{S5})\rangle$, i.e.,

$$\alpha \in Cn(R_{0a}, Sb(A_{S5})) \Rightarrow \Box\alpha \in Cn(R_{0a}, Sb(A_{S5})).$$

Without proof we state that the rule $r_\Box$ is derivable in $\langle R_{0a*}, A_{S5}\rangle$ (see [113], 1951):

Theorem 1.75. $\Box\alpha \in Cn(R_{0a*}, A_{S5} \cup \{\alpha\})$, *for every* $\alpha \in S_2$.

$S5$ was the name given by C.I. Lewis [60], 1932 to a system equivalent with $\langle R_{0a*}, A_{S5}\rangle$. In Kurt Gödel's paper [26], 1933, the term $S5$ had another meaning; there was considered a system equivalent with $\langle R_{0\Box*}, A_{S5}\rangle$, where $R_{0\Box*} = \{r_0, r_\Box, r_*\}$. Further in [59], 1957, $S5$ was in turn equivalent with $\langle R_{0*}, A_{S5}\rangle$. The above list of systems named as $S5$ is far from being complete (see, e.g., [23], 1965). It does not matter which rules are combined with a standard set of axioms as long as we speak only about formulas derivable in $S5$, since

$$Cn(R_{0*}, A_{S5}) = Cn(R_{0a*}, A_{S5}) = Cn(R_{0\Box*}, A_{S5}).$$

However, it can be shown that

$$\langle R_{0*}, A_{S5}\rangle \not\approx \langle R_{0a*}, A_{S5}\rangle \approx \langle R_{0\Box*}, A_{S5}\rangle$$

which means, in particular, that Gödel's rule $r_\Box$ is not derivable in $\langle R_{0*}, A_{S5}\rangle$ — see [135], 1982. Thus, derivability of rules as well as some metalogical properties, as for example the deduction theorem, are dependent upon the choice of primitive rules for $S5$. There is no difference in properties between $\langle R_{0a*}, A_{S5}\rangle$ and $\langle R_{0\Box*}, A_{S5}\rangle$ since both systems are equivalent (though $\langle R_{0\Box}, Sb(A_{S5})\rangle$ and $\langle R_{0\Box}, Sb(A_{S5})\rangle$ are not equivalent). But Meredith's version of $S5$, i.e., $\langle R_{0*}, A_{S5}\rangle$ is essentially weaker than those of Lewis and Gödel.

The fact that there are non-equivalent systems which have the same name $S5$ is apt to generate terminological confusion. To minimize such confusion the term $S5$ is used here only for $\langle R_{0a*}, A_{S5}\rangle$ (or $\langle R_{0a}, Sb(A_{S5})\rangle$ — the invariant version) of $S5$). The reader can clearly see that we have described the system $S5$ more extensively and more precisely than the remaining systems. It is so just because of the ambiguities mentioned above.

Łukasiewicz's logics

The many-valued logics of Łukasiewicz are usually defined by use of logical matrices — we take up this subject matter in the next chapter. Our approach to these systems is a bit non-standard as we define them syntactically. The ∞-valued *Łukasiewicz logic* is meant as the system $\langle R_{0*}, \text{Ł}_\infty\rangle$ (or $\langle R_0, Sb(\text{Ł}_\infty)\rangle$ — the invariant version) where the set Ł_∞ contains the following formulas:

(1) $(p \to q) \to ((q \to s) \to (p \to s))$

(2) $p \to (q \to p)$

(3) $((p \to q) \to q) \to ((q \to p) \to p)$

(4) $p \cdot q \to p$

(5) $p \cdot q \to q$

(6) $(p \to q) \to ((p \to s) \to (p \to q \cdot s))$

(7) $p \to p + q$

(8) $q \to p + q$

(9) $(p \to s) \to ((q \to s) \to (p + q \to s))$

(10) $(\sim p \to \sim q) \to (q \to p)$.

The n-valued logic of Łukasiewicz $\langle R_{0*}, \text{Ł}_n \rangle$ (or $\langle R_0, Sb(\text{Ł}_n) \rangle$), where $n \geqslant 2$, contains the above 10 axioms supplemented by the following two axioms (wherein we use the abbreviation $p \to^0 q = q$ and $p \to^{k+1} q = p \to (p \to^k q)$:

(11) $(p \to^n q) \to (p \to^{n-1} q)$

(12) $(p \equiv (p \to^k \sim p)) \to^{n-1} q$

for each $k \leqslant n$ such that $k + 2$ is not a divisor of $n - 1$. It is known that

$$Cn(R_{0*}, \text{Ł}_\infty) = \bigcap \{ Cn(R_{0*}, \text{Ł}_n) : n \geqslant 2 \}.$$

Łukasiewicz's many-valued logics possess many non-standard properties. For instance, we get the following variant of the deduction theorem, see [73], 1964;

Theorem 1.76. *For every* $X \subseteq S_2$ *and every* $\alpha, \beta \in S_2$:

(i) $\beta \in Cn(R_0, Sb(\text{Ł}_n) \cup X \cup \{\alpha\}) \Leftrightarrow (\alpha \to^{n-1} \beta) \in Cn(R_0, Sb(\text{Ł}_n) \cup X)$;

(ii) $\beta \in Cn(R_0, Sb(\text{Ł}_\infty) \cup X \cup \{\alpha\}) \Leftrightarrow \exists_n (\alpha \to^n \beta) \in Cn(R_0, Sb(\text{Ł}_\infty) \cup X)$.

Corollary 1.77. *For every* $X \subseteq S_2$ *and every* $\alpha \in S_2$:

(i) $Cn(R_0, Sb(\text{Ł}_n) \cup X \cup \{\alpha\}) = S_2 \Leftrightarrow (\alpha \to^{n-2} \sim \alpha) \in Cn(R_0, Sb(\text{Ł}_n) \cup X)$;

(ii) $Cn(R_0, Sb(\text{Ł}_\infty) \cup X \cup \{\alpha\}) = S_2 \Leftrightarrow \exists_n (\alpha \to^n \sim \alpha) \in Cn(R_0, Sb(\text{Ł}_\infty) \cup X)$.

Properties of the disjunction and conjunction in Łukasiewicz logics are pretty standard and quite similar to those of other systems, see Theorem 1.63, Corollary 1.67 and Theorem 1.71.

Chapter 2

Semantic methods in propositional logic

In the first chapter we have introduced syntactic notions concerning propositional logics. The purpose of the present chapter is to give a semantic approach to the further study of formal systems. This approach is algebraic in its nature and therefore we will use elementary notions and results of the theory of abstract algebra. Our discussion is based on the notion of the consequence operation generated by a given relational system. (Pre)ordered algebras are examined first and next we consider logical matrices. Then these structures are applied to define propositional logics. In Section 2.5 some relationships between propositional logics and lattice theory are presented.

2.1 Preordered sets

An attempt is made to adopt for (pre)ordered sets some concepts used in lattice theory. Special attention is given to the notion of a filter.

Preorderings

Let A be a non-empty set. A binary relation $\preccurlyeq$ on A is said to be a *preorder relation* (or *quasi-order* in the terminology due to H. Rasiowa and R. Sikorski [107], 1963) iff the following conditions hold for all $x, y, z \in A$:

(i) $x \preccurlyeq x$ (reflexivity),

(ii) $x \preccurlyeq y \land y \preccurlyeq z \implies x \preccurlyeq z$ (transitivity).

Any *order relation* fulfills, additionally, the condition

(iii) $x \preccurlyeq y \land y \preccurlyeq x \implies x = y$ (weak asymmetry).

A *preordered (ordered) set* is any pair $\langle A, \preccurlyeq \rangle$, where $\preccurlyeq$ is a preorder (order) relation on A.

Lemma 2.1. *Let $\langle A, \preccurlyeq \rangle$ be a preordered set. Then*

(i) *the relation $\approx$ defined on A as*

$$x \approx y \qquad \Leftrightarrow \qquad x \preccurlyeq y \ \wedge \ y \preccurlyeq x$$

    is an equivalence on A;

(ii) *the relation $\leqslant$ defined on $A/\approx$ by*

$$[x] \leqslant [y] \qquad \Leftrightarrow \qquad x \preccurlyeq y$$

    is an ordering on $A/\approx$.

Assume that $\langle A, \preccurlyeq \rangle$ is a preordered set and let $X \subseteq A$, $a \in A$. Then a is an *upper bound* of X, in symbols $a \in B_u(X)$, iff $y \preccurlyeq a$ for each $y \in X$. Similarly, a is called a *lower bound* of X, $a \in B_l(X)$, iff $a \preccurlyeq y$ for each $y \in X$. Moreover, we assume the following definitions:

(i) $\mathrm{Great}(X) = X \cap B_u(X)$,

(ii) $\mathrm{Least}(X) = X \cap B_l(X)$,

(iii) $\mathrm{Sup}(X) = \mathrm{Least}\left(B_u(X)\right)$,

(iv) $\mathrm{Inf}(X) = \mathrm{Great}\left(B_l(X)\right)$.

If $\preccurlyeq$ is an ordering, then the sets $\mathrm{Sup}(X), \mathrm{Inf}(X), \mathrm{Great}(X), \mathrm{Least}(X)$ contain at most one element — this element is the *supremum* of the set X, the *infimum* of the set X, the *greatest element* of X, the *least element* of X respectively. In case of preorderings, these sets may contain more than one element.

Lemma 2.2. *If $\langle A, \preccurlyeq \rangle$ is a preordered set and if $X, Y \subseteq A$, then*

(i) $X \subseteq Y \quad \Rightarrow \quad B_u(Y) \subseteq B_u(X) \ \wedge \ B_l(Y) \subseteq B_l(X)$;

(ii) $X \subseteq B_l\left(B_u(X)\right) \ \cap \ B_u\left(B_l(X)\right)$.

Lemma 2.3. *If $\langle A, \preccurlyeq \rangle$ is a preordered set, then*

(i) $B_u(\emptyset) = B_l(\emptyset) = A$;

(ii) $B_u(A) = \mathrm{Great}(A) \quad and \quad B_l(A) = \mathrm{Least}(A)$.

Both these lemmas are obvious.

Filters

If $\langle A, \preccurlyeq \rangle$ is a preordered set, then a non-empty set $H \subseteq A$ is a *filter* in $\langle A, \preccurlyeq \rangle$ iff $B_u\big(B_l(X)\big) \subseteq H$ for each $X \in \text{Fin}(H)$.

Our definition is not a literal translation of the usual definition of a filter in lattice theory. The reason is that we define the counterpart of the ordinary notion in preordered sets generally, not in a kind of pre-lattices. Moreover, we want to preserve somewhat more advanced properties of lattice filters whereas the faithful copy of the lattice definition, written down in Lemma 2.4, does not do it.

Lemma 2.4. *If H is a filter in $\langle A, \preccurlyeq \rangle$, then*

(i) $x \preccurlyeq y \,\wedge\, x \in H \,\Rightarrow\, y \in H$;

(ii) $X \in \text{Fin}(H) \,\Rightarrow\, \text{Inf}(X) \subseteq H$.

Easy proof of this lemma can be omitted. It should be emphasized, however, that from (i) and (ii) it does not follow that H is a filter in $\langle A, \preccurlyeq \rangle$. For instance, let $A = \{2, 3, 12, 18, 30\}$ and let

$$x \preccurlyeq y \quad \Leftrightarrow \quad x \text{ is a divisor of } y.$$

Then $\text{Inf}\{12, 18\} = \emptyset$, $\text{Inf}\{12\} = \{12\}$ and $\text{Inf}\{18\} = \{18\}$, hence the set $\{12, 18\}$ fulfills the conditions (i) and (ii) from Lemma 2.4. But $\{12, 18\}$ is not a filter since $B_u\big(B_l(\{12, 18\})\big) = \{12, 18, 30\}$.

Now we can prove some lemmas which show the soundness of the accepted definition of a filter.

Lemma 2.5. *If $\langle A, \leqslant \rangle$ is a lattice ordered set and if $H \subseteq A$ is non-empty, then H is a filter in $\langle A, \leqslant \rangle$ iff*

(i) $x \leqslant y \,\wedge\, x \in H \,\Rightarrow\, y \in H$;

(ii) $x, y \in H \,\Rightarrow\, x \cap y \in H$.

Proof. Let H be a lattice filter (i.e., fulfills (i) and (ii) above), $X \in \text{Fin}(H)$ and $a \in B_u\big(B_l(X)\big)$. Then $x \leqslant a$ for every $x \in B_l(X)$, hence $a_1 \cap \ldots \cap a_n \leqslant a$ where $X = \{a_1, \ldots, a_n\}$. But $a_1 \cap \ldots \cap a_n \in H$ by (ii), thus $a \in H$ by (i).

Let H be a filter in $\langle A, \leqslant \rangle$, By Lemma 2.4 (i) we need to prove (ii) only. Assume that $x, y \in H$. We have, of course, $x \cap y \in \text{Inf}\{x, y\}$ for every $x, y \in A$. Then, by 2.4 (ii), we get $x \cap y \in \text{Inf}\{x, y\} \subseteq H$. $\square$

Lemma 2.6. *If $\langle A, \preccurlyeq \rangle$ is a preordered set, then $B_u\big(B_l(X)\big)$ is a filter in $\langle A, \preccurlyeq \rangle$ for every non-empty $X \subseteq A$.*

Proof. Let X_0 be a finite subset of $B_u\big(B_l(X)\big)$. By Lemma 2.2 (i),

$$B_u\big(B_l(X_0)\big) \subseteq B_u\big(B_l\big(B_u\big(B_l(X)\big)\big)\big)$$

and by 2.2 (ii),

$$B_l(X) \subseteq B_l\big(B_u\big(B_l(X)\big)\big).$$

Thus, by Lemma 2.2 (i), $B_u\big(B_l(X_0)\big) \subseteq B_u\big(B_l(X)\big)$. $\square$

Since $B_u(\{x\}) = B_u(B_l(\{x\}))$ for every $x \in A$, we obtain from Lemma 2.6:

Corollary 2.7. *If $\langle A, \preccurlyeq \rangle$ is a preordered set, then $B_u(\{x\})$ is a filter in $\langle A, \preccurlyeq \rangle$ for every $x \in A$.*

The set $B_u(\{x\}) = \{z \in A : x \preccurlyeq z\}$ will be called the *principal filter* generated by the element x.

Lemma 2.8. *If $\langle A, \preccurlyeq \rangle$ is a preordered set, then*

(i) *the intersection of any family of filters in $\langle A, \preccurlyeq \rangle$ is a filter provided it is not empty;*

(ii) *the union of any (non-empty) chain of filters is a filter.*

Proof. (i): Assume that $\mathscr{L}$ is a family of filters in $\langle A, \preccurlyeq \rangle$ such that $\bigcap \mathscr{L} \neq \emptyset$ and let X be a finite subset of $\bigcap \mathscr{L}$. Then $X \subseteq H$ for each $H \in \mathscr{L}$ and, since H is a filter, $B_u(B_l(X)) \subseteq H$.

Thus, $B_u(B_l(X)) \subseteq \bigcap\{H : H \in \mathscr{L}\} = \bigcap \mathscr{L}$ which was to be proved.

(ii): Let $\mathscr{L}$ be a chain of filters in $\langle A, \preccurlyeq \rangle$. Then, for every finite $X \subseteq \bigcup \mathscr{L}$ there exists $H \in \mathscr{L}$ such that $X \subseteq H$. Thus $B_u(B_l(X)) \subseteq H \subseteq \bigcup \mathscr{L}$, hence $\bigcup \mathscr{L}$ is a filter. □

The family of all filters in $\langle A, \preccurlyeq \rangle$ can be regarded as an ordered set with the order relation being the set-theoretical inclusion. Let us prove a lemma from which it immediately follows that $\langle A, \preccurlyeq \rangle$ need not contain the least filter.

Lemma 2.9. *In any preordered set $\langle A, \preccurlyeq \rangle$,*

$$\mathrm{Great}(A) = \bigcap\{H : H \text{ is a filter in } A\}.$$

Proof. We have $B_u(B_l(\emptyset)) = \mathrm{Great}(A)$ by Lemma 2.3. Hence $\mathrm{Great}(A) \subseteq H$ if H is a filter.

To prove the inclusion ($\supseteq$) it suffices to consider all principal filters, i.e., the sets $B_u(\{x\})$ for $x \in A$, see Corollary 2.7. If y is an element of all filters, then $y \in B_u(\{x\})$ for each $x \in A$. Thus $x \preccurlyeq y$ for each $x \in A$ and hence $y \in \mathrm{Great}(A)$ by the definition. □

The family of all filters in $\langle A, \preccurlyeq \rangle$ contains the least element iff $\mathrm{Great}(A) \neq \emptyset$ and this least element, if it exists, is equal to $\mathrm{Great}(A)$. A filter H in $\langle A, \preccurlyeq \rangle$ is said to be *proper* provided that $H \neq A$. A proper filter is called *maximal* iff it is a maximal element in the family of all proper filters.

Theorem 2.10. *If $\langle A, \preccurlyeq \rangle$ is a preordered set and $\mathrm{Least}(A) \neq \emptyset$, then any proper filter is contained in a maximal filter.*

An easy proof based on Zorn's lemma will be omitted.

A general definition of an ideal in ordered sets was given by [24], 1954. The dual definition of a filter can be found in R. Suszko [116], 1977. However, that notion is not equivalent to the just defined notion of a filter in a preordered set; these two notions coincide only in lattices.

Filters generated by sets

Let $\langle A, \preccurlyeq \rangle$ be a preordered set. A filter H in $\langle A, \preccurlyeq \rangle$ is called the *filter generated by* a (non-empty) set $X \subseteq A$ if H is the least, with respect to inclusion, filter containing X. Since the family of all filters is closed under arbitrary non-empty intersections (see Lemma 2.8 (i)), the filter generated by X can be constructed as the intersection of all filters containing X.

Let us define the filter operation $F : 2^A \to 2^A$. Take $X \subseteq A$ and set

$$F(X) = \bigcap \{H : H \text{ is a filter containing } X\}.$$

If X is non-empty, then obviously $F(X)$ is the filter generated by X. It is easy to prove

Corollary 2.11. *For every non-empty $H \subseteq A$, $H \neq A$:*

(i) *H is a filter iff $F(H) = H$;*

(ii) *H is a maximal filter iff $H = F(H)$ and $F(H \cup \{x\}) = A$ for every $x \notin H$.*

The next result is an easy consequence of the introduced definitions.

Corollary 2.12. *For every $X, Y \subseteq A$:*

(i) *$X \subseteq F(X)$;*

(ii) *$X \subseteq Y \Rightarrow F(X) \subseteq F(Y)$;*

(iii) *$F\big(F(X)\big) \subseteq F(X)$;*

(iv) *$F(X) = \bigcup \{F(Y) : Y \in \text{Fin}(X)\}$.*

Lemma 2.13. *In every preordered set $\langle A, \preccurlyeq \rangle$:*

(i) *$F(\emptyset) = F\big(\text{Great}(A)\big) = \text{Great}(A) = B_u\big(B_l(\emptyset)\big)$;*

(ii) *$F(X) = B_u\big(B_l(X)\big)$, for each finite set $X \subseteq A$;*

(iii) *$F(\{a\}) = A$ if $a \in \text{Least}(A)$.*

Proof. (i): By Lemma 2.9 and Lemma 2.3 we get immediately

$$F(\emptyset) = \text{Great}(A) = B_u\big(B_l(\emptyset)\big).$$

Moreover, it immediately follows from Corollary 2.12 that $F\big(F(\emptyset)\big) = F(\emptyset)$ and hence $F\big(\text{Great}(A)\big) = F\big(F(\emptyset)\big) = F(\emptyset)$.

(ii): The case that X is empty is proved by (i). Assume that $\emptyset \neq X \in \text{Fin}(A)$. Then $B_u\big(B_l(X)\big) \subseteq F(X)$, since $X \subseteq F(X)$ and $F(X)$ is a filter. On the other hand, $F(X) \subseteq B_u\big(B_l(X)\big)$ by Lemma 2.6.

(iii): Let us assume that $a \in \text{Least}(A)$. Then $a \preccurlyeq x$ for all $x \in A$. The set $F(\{a\})$ is a filter, thus it follows from Lemma 2.4 (i) that $x \in F(\{a\})$ for all $x \in A$. $\qquad \square$

Induced preorderings

Given a preordered set $\langle A, \preccurlyeq \rangle$ and a non-empty subset $B \subseteq A$, we can induce on B the relation $\preccurlyeq_B$ by

$$x \preccurlyeq_B y \qquad \Leftrightarrow \qquad x \preccurlyeq y, \qquad \text{for all } x, y \in B.$$

The relation $\preccurlyeq_B$ is called the restriction of $\preccurlyeq$ to the set B. If there is no danger of confusion we shall omit the subscript B. Note that the restriction of a preorder relation is always a preorder relation.

Lemma 2.14. *If $\langle A, \preccurlyeq \rangle$ is a preordered set and if $\emptyset \neq B \subseteq A$, then for each filter H in $\langle B, \preccurlyeq_B \rangle$ there is a filter G in $\langle A, \preccurlyeq \rangle$ such that*

$$H = G \cap B.$$

Proof. Assume that H is a filter in $\langle B, \preccurlyeq_B \rangle$ and let G be the filter in $\langle A, \preccurlyeq \rangle$ generated by the set H. Obviously $H \subseteq G \cap B$ and we have to prove only that $G \cap B \subseteq H$.

Let us agree that $B_l(X)$ and $B_u(X)$, for $X \subseteq A$, will denote the sets of bounds in $\langle A, \preccurlyeq \rangle$ and that $B_l^1(X)$, $B_u^1(X)$ (for $X \subseteq B$) will be the subsets of B determined by the relation $\preccurlyeq_B$.

It is easy to verify that

$$B_l^1(X) = B \cap B_l(X) \ \text{ and } \ B_u^1(X) = B \cap B_u(X)$$

for each $X \subseteq B$.

Suppose that $x \in G \cap B$. Since G is the filter in $\langle A, \preccurlyeq \rangle$ generated by H, it follows from Corollary 2.12 (iv) and Lemma 2.13 (ii) that $x \in B \cap B_u\big(B_l(X)\big)$ for some finite set $X \subseteq H$. By Lemma 2.2 (i),

$$B_u\big(B_l(X)\big) \subseteq B_u\big(B \cap B_l(X)\big)$$

and hence

$$x \in B \cap \big(B_u\big(B \cap B_l(X)\big)\big) = B_u^1\big(B_l^1(X)\big).$$

But H is a filter in $\langle B, \preccurlyeq_B \rangle$ and $X \in \mathrm{Fin}(H)$, then $x \in H$. $\qquad\qquad\square$

Let us note that there is a preordered set $\langle A, \preccurlyeq \rangle$ such that $G \cap B$ is not a filter in $\langle B, \preccurlyeq_B \rangle$ for some filter G in $\langle A, \preccurlyeq \rangle$ and non-empty $B \subseteq A$. For instance, let $A = \{2, 4, 8, 12, 48\}$ and let

$$x \preccurlyeq y \qquad \Leftrightarrow \qquad x \text{ is a divisor of } y.$$

Then $G = \{4, 8, 12, 48\}$ is a filter in $\langle A, \preccurlyeq \rangle$ but $G \cap B$ is not a filter in $\langle B, \preccurlyeq \rangle$ where $B = \{2, 8, 12, 48\}$.

Products of preordered sets

Let $\{\langle A_t, \preceq_t \rangle\}_{t \in T}$ be an indexed family of preordered sets. The product of this family will be considered as the pair

$$\langle \mathbf{P}_{t \in T} A_t, \preceq \rangle$$

where the preordering $\preceq$ is defined as

$$\langle x_t \rangle_{t \in T} \preceq \langle y_t \rangle_{t \in T} \quad \Leftrightarrow \quad \left(x_t \preceq_t y_t, \text{ for each } t \in T \right).$$

The relation $\preceq$ can also be defined by

$$x \preceq y \quad \Leftrightarrow \quad \left(\pi_t(x) \preceq_t \pi_t(y), \text{ for each } t \in T \right)$$

where x, y are elements of the product and π_t is the projection onto the t-axis.

The symbols $B_u^t(\ )$, $B_l^t(\ )$, $\mathrm{Great}^t(\ ), \ldots$ stand for the subsets of A_t in relation to $\preceq_t$. The appropriate subsets of the product will be denoted by $B_u(\ )$, $B_l(\ )$, $\mathrm{Great}(\ ), \ldots$ and so on.

Let $X \subseteq \mathbf{P}_{t \in T} A_t$ be a set which is kept fixed in the next three statements.

Lemma 2.15.

(i) $B_u(X) = \mathbf{P}_{t \in T} B_u^t \left(\pi_t(X) \right)$;

(ii) $B_l(X) = \mathbf{P}_{t \in T} B_l^t \left(\pi_t(X) \right)$.

Proof. (i): If $a \in B_u(X)$, then $\pi_t(x) \preceq_t \pi_t(a)$ for every $x \in X$ and every $t \in T$, hence $\pi_t(a) \in B_u^t \left(\pi_t(X) \right)$ for all $t \in T$.

On the other hand, if $a \in \mathbf{P}_{t \in T} B_u^t \left(\pi_t(X) \right)$, then $\pi_t(a) \in B_u^t \left(\pi_t(X) \right)$ for each $t \in T$. Thus $\pi_t(x) \preceq_t \pi_t(a)$ for all $t \in T$ and for all $x \in X$, therefore $a \in B_u(X)$.

The proof of (ii) is similar. $\qquad\square$

Corollary 2.16.

(i) $\pi_t \left(B_u(X) \right) = B_u^t \left(\pi_t(X) \right)$, *if* $B_u(X) \neq \emptyset$;

(ii) $\pi_t \left(B_l(X) \right) = B_l^t \left(\pi_t(X) \right)$, *if* $B_l(X) \neq \emptyset$.

Lemma 2.17.

(i) $\mathrm{Great}(X) = X \cap \mathbf{P}_{t \in T} \mathrm{Great}^t \left(\pi_t(X) \right)$;

(ii) $\mathrm{Least}(X) = X \cap \mathbf{P}_{t \in T} \mathrm{Least}^t \left(\pi_t(X) \right)$;

(iii) $\mathrm{Sup}(X) = \mathbf{P}_{t \in T} \mathrm{Sup}^t \left(\pi_t(X) \right)$;

(iv) $\mathrm{Inf}(X) = \mathbf{P}_{t \in T} \mathrm{Inf}^t \left(\pi_t(X) \right)$.

Proof. It will be only shown that (i), (iv) hold because the proofs of (ii), (iii) are quite similar.

(i): If $a \in \text{Great}(X) = X \cap B_u(X)$, then if follows from Corollary 2.16 that $\pi_t(a) \in B_u^t(\pi_t(X))$ for every $t \in T$ and hence $\pi_t(a) \in \pi_t(X) \cap B_u^t(\pi_t(X)) = \text{Great}^t(\pi_t(X))$ for every $t \in T$. On the other hand,

$$X \cap \mathbf{P}_{t \in T} \text{Great}^t(\pi_t(X)) \subseteq X \cap \mathbf{P}_{t \in T} B_u^t(\pi_t(X)) = X \cap B_u(X) = \text{Great}(X).$$

(iv): We have

$$\text{Inf}(X) = \text{Great}(B_l(X)) = B_l(X) \cap \mathbf{P}_{t \in T} \text{Great}^t(\pi_t(B_l(X)))$$

$$\overset{2.16}{=} B_l(X) \cap \mathbf{P}_{t \in T} \text{Great}^t(B_l^t(\pi_t(X))) = \mathbf{P}_{t \in T} B_l^t(\pi_t(X)) \cap \mathbf{P}_{t \in T} \text{Inf}^t(\pi_t(X))$$

$$= \mathbf{P}_{t \in T}(B_l^t(\pi_t(X)) \cap \text{Inf}^t(\pi_t(X))) = \mathbf{P}_{t \in T} \text{Inf}^t(\pi_t(X)). \qquad \square$$

The following theorem describes some properties of filters in products.

Theorem 2.18. *Assume that $B_l^t(Y) \neq \emptyset$ for all $t \in T$ and all finite $Y \subseteq A_t$. Then:*

(i) *if H_t, for each $t \in T$, is a filter in $\langle A, \preccurlyeq_t \rangle$, then $\mathbf{P}_{t \in T} H_t$ is a filter in the product;*

(ii) *if H is a filter in the product, then $\pi_t(H)$ is a filter in $\langle A_t, \preccurlyeq_t \rangle$.*

Proof. (i): Let X be a finite subset of $\mathbf{P}_{t \in T} H_t$. Then, obviously, $\pi_t(X) \in \text{Fin}(H_t)$ for each $t \in T$ and $B_u^t(B_l^t(\pi_t(X))) \subseteq H_t$ as H_t is a filter. Hence by Lemma 2.15 and Corollary 2.16,

$$B_u(B_l(X)) = \mathbf{P}_{t \in T} B_u^t(B_l^t(\pi_t(X))) \subseteq \mathbf{P}_{t \in T} H_t.$$

(ii): Let X_0 be a finite subset of $\pi_t(H)$ for some $t \in T$. Then there exists $X \in \text{Fin}(H)$ such that $X_0 = \pi_t(X)$. Since H is a filter, we obtain

$$\mathbf{P}_{s \in T} B_u^s(B_l^s(\pi_s(X))) = B_u(B_l(X)) \subseteq H$$

on the basis of 2.15 and 2.16 and hence

$$B_u^t(B_l^t(X_0)) = B_u^t(B_l^t(\pi_t(X))) \subseteq \pi_t(H). \qquad \square$$

It should be remarked that filters in the product need not be of the form $\mathbf{P}_{t \in T} H_t$ where H_t, for any $t \in T$, is a filter in A_t. For instance, let T be an infinite set and $A_t = \{0, 1\}$ for every $t \in T$ and the ordering $\preccurlyeq_t$ is induced from the set of natural numbers, i.e., $0 \preccurlyeq_t 1$, $0 \preccurlyeq_t 0$, $1 \preccurlyeq_t 1$. Take

$$H = \{x : \pi_t(x) = 0 \text{ for a finite number of } t\text{'s }\}$$

and note that H is a filter in the product. Moreover, $\pi_t(H) = \{0,1\}$ for each $t \in T$ and hence $H \neq \mathop{\mathbf{P}}\limits_{t \in T} A_t = \mathop{\mathbf{P}}\limits_{t \in T} \pi_t(H)$.

According to our convention the symbol $F^t(\ )$ stands for the filter operation in $\langle A_t, \preccurlyeq_t \rangle$; the operation in the product is denoted by $\mathbf{F}(\ )$.

Theorem 2.19. *Assume that $B_l^t(Y) \neq \emptyset$ for all $t \in T$ and all finite $Y \subseteq A_t$. Then*

(i) $\mathbf{F}(X) \subseteq \mathop{\mathbf{P}}\limits_{t \in T} F^t\big(\pi_t(X)\big)$, *for every $X \subseteq \mathop{\mathbf{P}}\limits_{t \in T} A_t$;*

(ii) $\mathbf{F}(X) = \mathop{\mathbf{P}}\limits_{t \in T} F^t\big(\pi_t(X)\big)$, *for every finite $X \subseteq \mathop{\mathbf{P}}\limits_{t \in T} A_t$.*

Proof. (i): If X is empty, then according to Lemma 2.13 (i) we need to show

$$\text{Great}\Big(\mathop{\mathbf{P}}\limits_{t \in T} A_t\Big) \subseteq \mathop{\mathbf{P}}\limits_{t \in T} \text{Great}^t(A_t).$$

This inclusion, however, follows immediately from Lemma 2.17 (i).

Let $X \neq \emptyset$. Note that $X \subseteq \mathop{\mathbf{P}}\limits_{t \in T} \pi_t(X) \subseteq \mathop{\mathbf{P}}\limits_{t \in T} F^t\big(\pi_t(X)\big)$. Moreover, by Theorem 2.18 (i), the set $\mathop{\mathbf{P}}\limits_{t \in T} F^t\big(\pi_t(X)\big)$ is a filter in the product. Thus $\mathbf{F}(X) \subseteq \mathop{\mathbf{P}}\limits_{t \in T} F^t\big(\pi_t(X)\big)$ since $\mathbf{F}(X)$ is the least filter in the product containing X.

(ii): If X is finite, then $\pi_t(X) \in \text{Fin}(A_t)$ for every $t \in T$. Hence, by Lemma 2.13 (ii), $F^t\big(\pi_t(X)\big) = B_u^t\big(B_l^t(\pi_t(X))\big)$ and $\mathbf{F}(X) = B_u\big(B_l(X)\big)$. Thus, it follows from Lemma 2.15 and Corollary 2.16 that $\mathbf{F}(X) = B_u\big(B_l(X)\big) = \mathop{\mathbf{P}}\limits_{t \in T} B_u^t\big(\pi_t(B_l(X))\big) = \mathop{\mathbf{P}}\limits_{t \in T} B_u^t\big(B_l^t(\pi_t(X))\big) = \mathop{\mathbf{P}}\limits_{t \in T} F^t\big(\pi_t(X)\big)$. $\quad\square$

The next result gives some further information on filters in finite products.

Theorem 2.20. *Assume that the set T is finite and let $B_l^t(Y) \neq \emptyset$ for all $t \in T$ and for all finite $Y \subseteq A_t$. Then H is a filter in the product if and only if for every $t \in T$ there is a filter H_t in $\langle A_t, \preccurlyeq_t \rangle$ such that $H = \mathop{\mathbf{P}}\limits_{t \in T} H_t$.*

Proof. By Theorem 2.18 it suffices to show that

$$H = \mathop{\mathbf{P}}\limits_{t \in T} \pi_t(H)$$

for each filter H in the product.

The inclusion $(\subseteq)$ is obvious. To prove $(\supseteq)$ assume that $\langle x_t \rangle_{t \in T} \in \mathop{\mathbf{P}}\limits_{t \in T} \pi_t(H)$. Since the set T is finite, there exists then a finite set $X \subseteq H$ such that $x_t \in \pi_t(X)$ for every $t \in T$.

The set H is a filter, hence $B_u\big(B_l(X)\big) \subseteq H$. Thus, by Lemma 2.15 and Corollary 2.16,

$$\langle x_t \rangle_{t \in T} \in \mathop{\mathbf{P}}\limits_{t \in T} \pi_t(X) \subseteq \mathop{\mathbf{P}}\limits_{t \in T} B_u^t\big(B_l^t(\pi_t(X))\big)$$

$$\subseteq \mathop{\mathbf{P}}\limits_{t \in T} B_u^t\big(\pi_t(B_l(X))\big) \subseteq B_u\big(B_l(X)\big) \subseteq H. \quad\square$$

Corollary 2.21. *If T is finite, then $\mathbf{F}(X) = \mathop{\mathbf{P}}\limits_{t \in T} F^t\big(\pi_t(X)\big)$ for every set $X \subseteq \mathop{\mathbf{P}}\limits_{t \in T} A_t$, provided that $B_l^t(Y) \neq \emptyset$ for all $t \in T$ and all finite $Y \subseteq A_t$.*

2.2   Preordered algebras

A preordered algebra is a system $\langle \mathscr{A}, \preccurlyeq \rangle$ where $\mathscr{A}$ is an abstract algebra and $\preccurlyeq$ is a preorder relation on A (as we agreed A denotes the universe of $\mathscr{A}$). Preordered algebras will be denoted by German capital letters $\mathfrak{A}$, $\mathfrak{B}$, $\mathfrak{C}$, ...

As an example of the notion let us mention any system $\langle A, \dot{\rightarrow}, \cup, \cap, -, \leqslant \rangle$ where $\dot{\rightarrow}, \cup, \cap, -$ are Heyting (or Boolean) operations defined in the standard way by means of the order relation $\leqslant$ on A. Let us note that any Heyting order relation $\leqslant$ can also be defined by use of $\dot{\rightarrow}, \cup, \cap, -$. In the general case of a preordered (or ordered) algebra, any relationships between the operations and the preordering need not appear.

Consequence operations

Let $\mathscr{S}$ be the algebra of a fixed propositional language based on the infinite set At of propositional variables and let $\mathfrak{A} = \langle \mathscr{A}, \preccurlyeq \rangle$ be a preordered algebra such that $\mathscr{A}$ and $\mathscr{S}$ are similar. We define in a standard way (see [115], 1962, [86], 1974) the consequence operation generated by $\mathfrak{A}$:

$$\overrightarrow{\mathfrak{A}} : 2^S \rightarrow 2^S.$$

Definition 2.22. For every $X \subseteq S$ and every $\alpha \in S$,

$$\alpha \in \overrightarrow{\mathfrak{A}}(X) \quad \Leftrightarrow \quad \left(h^v(\alpha) \in F\left(h^v(X) \right), \text{ for every } v \colon At \rightarrow A \right)$$

where $F \colon 2^A \rightarrow 2^A$ is the filter operation determined by the preordering $\preccurlyeq$ and $h^v \colon \mathscr{S} \rightarrow \mathscr{A}$ is the homomorphism generated by the valuation $v \colon At \rightarrow A$.

Since $F\left(h^v(X) \right)$ is the least filter, provided it is non-empty, containing $h^v(X)$, we can equivalently reformulate Definition 2.22 as follows.

Lemma 2.23. $\alpha \in \overrightarrow{\mathfrak{A}}(X)$ *if and only if* $h^v(X) \subseteq H \Rightarrow h^v(\alpha) \in H$ *for every filter H in $\langle A, \preccurlyeq \rangle$ and every valuation $v \colon At \rightarrow A$.*

As an immediate result of 2.22 and 2.12 we obtain

Corollary 2.24. *For every $X, Y \subseteq S$:*

(i) $X \subseteq \overrightarrow{\mathfrak{A}}(X)$;

(ii) $X \subseteq Y \quad \Rightarrow \quad \overrightarrow{\mathfrak{A}}(X) \subseteq \overrightarrow{\mathfrak{A}}(Y)$;

(iii) $\overrightarrow{\mathfrak{A}}\left(\overrightarrow{\mathfrak{A}}(X) \right) \subseteq \overrightarrow{\mathfrak{A}}(X)$;

(iv) $h^e\left(\overrightarrow{\mathfrak{A}}(X) \right) \subseteq \overrightarrow{\mathfrak{A}}\left(h^e(X) \right)$, *for every $e \colon At \rightarrow S$.*

Thus, $\overrightarrow{\mathfrak{A}}$ is a structural consequence operation over S. Let us note that $\overrightarrow{\mathfrak{A}}$ need not be finitistic. Some examples of non-finitistic operations $\overrightarrow{\mathfrak{A}}$ are presented in Section 2.5.

Next we define the set $E(\mathfrak{A})$ of all formulas valid (or true) in $\mathfrak{A}$.

Definition 2.25. $\quad \alpha \in E(\mathfrak{A}) \quad \Leftrightarrow \quad (h^v(\alpha) \in \mathrm{Great}(A), \text{ for every } v \colon At \to A)$.

Since $\mathrm{Great}(A) = F(\emptyset)$ — see Lemma 2.13 (i) — we conclude that a formula α is $\mathfrak{A}$-valid if and only if α is derivable by means of $\overrightarrow{\mathfrak{A}}$ from the empty set, i.e.,

$$E(\mathfrak{A}) = \overrightarrow{\mathfrak{A}}(\emptyset).$$

Let $\mathfrak{A} = \langle \mathscr{A}, \preccurlyeq \rangle$ and $\mathfrak{B} = \langle \mathscr{B}, \preccurlyeq_1 \rangle$ be two similar preordered algebras. If there is an isomorphism h between the algebras $\mathscr{A}$ and $\mathscr{B}$ such that

$$x \preccurlyeq y \quad \Leftrightarrow \quad h(x) \preccurlyeq_1 h(y), \qquad \text{for all } x, y \in A,$$

then $\mathfrak{A}$ and $\mathfrak{B}$ are called isomorphic. We will write $\mathfrak{A} \cong \mathfrak{B}$ for '$\mathfrak{A}$ and $\mathfrak{B}$ are isomorphic'. Further we will examine only those properties of preordered algebras which are invariant under isomorphism. Therefore, we will not distinguish isomorphic structures.

$\mathfrak{B} = \langle \mathscr{B}, \preccurlyeq_1 \rangle$ is called a substructure of $\mathfrak{A} = \langle \mathscr{A}, \preccurlyeq \rangle$, in symbols $\mathfrak{B} \subseteq \mathfrak{A}$, iff $\mathscr{B}$ is a subalgebra of $\mathscr{A}$ and $\preccurlyeq_1$ is the restriction of $\preccurlyeq$ to $\mathfrak{B}$.

Lemma 2.26. *For every pair of preordered algebras $\mathfrak{A}$ and $\mathfrak{B}$:*

(i) *if $\mathfrak{A} \cong \mathfrak{B}$, then $\overrightarrow{\mathfrak{A}} = \overrightarrow{\mathfrak{B}}$;*

(ii) *if $\mathfrak{B} \subseteq \mathfrak{A}$, then $\overrightarrow{\mathfrak{A}} \leqslant \overrightarrow{\mathfrak{B}}$.*

Proof. The property (i) is obvious. To show (ii) let us assume that $\mathfrak{B} \subseteq \mathfrak{A}$ and let $\alpha \notin \overrightarrow{\mathfrak{B}}(X)$ for some $X \subseteq S$, $\alpha \in S$.

From Lemma 2.23 it follows that $h^v(X) \subseteq H$ and $h^v(\alpha) \notin H$ for some valuation $v \colon At \to B$ and some filter H in $\mathfrak{B}$. But the preordering in $\mathfrak{B}$ is induced from $\mathfrak{A}$ and hence, by Lemma 2.14, $H = B \cap G$ for some filter G in $\mathfrak{A}$. We have $h^v(X) \subseteq B \cap G$ and $h^v(\alpha) \in B \setminus H$. Thus, $h^v(X) \subseteq G$ and $h^v(\alpha) \notin G$ which yields, on the basis of 2.23, $\alpha \notin \overrightarrow{\mathfrak{A}}(X)$. $\qquad \square$

Finite preordered algebras

As it was mentioned, the operation $\overrightarrow{\mathfrak{A}}$ need not be finitistic. It can be shown, however, that finite preordered algebras generate finitistic consequence operations.

Theorem 2.27. *If $\mathfrak{A} = \langle \mathscr{A}, \preccurlyeq \rangle$ is a finite preordered algebra and $X \subseteq S$, then*

$$\overrightarrow{\mathfrak{A}}(X) \subseteq \bigcup \{ \overrightarrow{\mathfrak{A}}(Y) : Y \in \mathrm{Fin}(X) \}.$$

Proof. Assume that $\alpha \notin \overrightarrow{\mathfrak{A}}(Y)$ for each $Y \in \mathrm{Fin}(X)$. The symbol V stands for the family of all valuations $v \colon At \to A$ and let $V(Y)$, for each $Y \in \mathrm{Fin}(X)$, be defined by

(i) $\quad v \in V(Y) \quad \Leftrightarrow \quad h^v(\alpha) \notin F\big(h^v(Y)\big).$

From the assumptions it immediately follows that $V(Y)$ is non-empty for each $Y \in \mathrm{Fin}(X)$. Moreover, it is easy to observe that

(ii) $Y_1 \subseteq Y_2 \ \Rightarrow \ V(Y_2) \subseteq V(Y_1)$,   for all $Y_1, Y_2 \in \mathrm{Fin}(X)$.

We will consider the Boolean algebra determined by $\langle 2^V, \subseteq \rangle$ where $\subseteq$ is the set-theoretical relation of inclusion. By (ii), we have $\emptyset \neq V(Y_1 \cup \ldots \cup Y_n) \subseteq V(Y_1) \cap \ldots \cap V(Y_n)$ and hence the set $V(Y_1) \cap \ldots \cap V(Y_n)$ is non-empty for every $Y_1, \ldots, Y_n \in \mathrm{Fin}(X)$. Thus, the subset $\{V(Y) : Y \in \mathrm{Fin}(X)\}$ of the Boolean algebra $\langle 2^V, \subseteq \rangle$ has the finite intersection property. Then there exists an ultrafilter H in 2^V such that $V(Y) \in H$ for each finite set $Y \subseteq X$ (see Theorem 1.18). Of course, $V \in H$ because V is the greatest element in 2^V. Moreover,

(iii) $\bigcup_{a \in A} \{v \in V : v(p) = a\} = V$, for every $p \in At$.

But H is a prime filter — see Corollary 1.22 — and A is a finite set, hence by (iii) we get

(iv) For every $p \in At$ there is an element $a \in A$ such that

$$\{v \in V : v(p) = a\} \in H.$$

Moreover, it is obvious that

(v) $\{v \in V : v(p) = a\} \cap \{v \in V : v(p) = b\} = \emptyset \notin H$ if $a \neq b$.

Let us define a valuation $w \colon At \to A$ by putting

(vi) $w(p) = a \ \Leftrightarrow \ \{v \in V : v(p) = a\} \in H$, for every $p \in At$.

From (iv) and (v) it follows that the definition is correct. Obviously,

(vii) $\{v \in V : v(p) = w(p)\} \in H$, for every $p \in At$.

Consider $Y \in \mathrm{Fin}(X)$ and let $M = At(Y) \cup At(\alpha)$. Since M is finite and H is a filter, by (vii) we get

(viii) $\{v \in V : v|_M = w|_M\} = \bigcap_{p \in M} \{v \in V : v(p) = w(p)\} \in H$.

Thus, we conclude that the set $V(Y) \cap \{v \in V : v|_M = w|_M\}$ is non-empty as an element of H. Then, by (i), there is a $v \in V$ such that $v|_M = w|_M$ and $h^v(\alpha) \notin F(h^v(Y))$. But it has been assumed that $M = At(Y) \cup At(\alpha)$, hence $h^w(\alpha) \notin F(h^w(Y))$ for any finite set $Y \subseteq X$. By Corollary 2.12 (iv), $h^w(\alpha) \notin F(h^w(X))$, which completes the proof. $\square$

The above theorem is a version of the known result by J. Łoś and R. Suszko [64], 1958. In the above proof we have used the technique of ultraproducts.

$\mathfrak{A}$-valid and $\mathfrak{A}$-normal rules

In the previous chapter we defined admissible and derivable rules of a given syntactical system $\langle R, X \rangle$, see Definition 1.45. If we turn to semantic treatment of propositional logic the situation changes. We can admittedly take an arbitrary set $X \subseteq S$ as the set of axioms (or premises) and the set of primitive rules is replaced by the consequence operation $\overrightarrow{\mathfrak{A}}$ generated by a given semantics $\mathfrak{A}$. Now, the counterparts of admissible and derivable rules, namely the *valid* (unfailing) and *normal rules*, are defined as follows:

Definition 2.28. For each $X \subseteq S$ and $r \in \mathscr{R}_S$:

(i) $r \in V(\mathfrak{A}, X) \quad \Leftrightarrow \quad (\Pi \subseteq \overrightarrow{\mathfrak{A}}(X) \Rightarrow \alpha \in \overrightarrow{\mathfrak{A}}(X), \text{ for each } \langle \Pi, \alpha \rangle \in r)$;

(ii) $r \in N(\mathfrak{A}, X) \quad \Leftrightarrow \quad \alpha \in \overrightarrow{\mathfrak{A}}(X \cup \Pi), \quad \text{for each } \langle \Pi, \alpha \rangle \in r$.

If $X = \emptyset$, then instead of $V(\mathfrak{A}, \emptyset)$ (or $N(\mathfrak{A}, \emptyset)$) we will write down $V(\mathfrak{A})$ (or $N(\mathfrak{A})$). The rule r is $\mathfrak{A}$-valid iff $r \in V(\mathfrak{A})$, and r is $\mathfrak{A}$-normal iff $r \in N(\mathfrak{A})$. Of course, $V(\mathfrak{A}, X) = \mathrm{Adm}(N(\mathfrak{A}), X) = \mathrm{Adm}(V(\mathfrak{A}, X), X)$ and $N(\mathfrak{A}, X) = \mathrm{Der}(N(\mathfrak{A}), X)$ for each set X. The connections between validity and normality of rules can be stated by the equations

$$N(\mathfrak{A}, X) = \bigcap \{V(\mathfrak{A}, Y) : X \subseteq Y\},$$

$$N(\mathfrak{A}) = \bigcap \{V(\mathfrak{A}, X) : X \subseteq S\}.$$

The analogy between admissible and valid rules as well as between derivable and normal rules is quite clear. It is obvious that

$$N(\mathfrak{A}) = \mathrm{Der}(\overrightarrow{\mathfrak{A}}) = \mathrm{Der}\big(N(\mathfrak{A}), E(\mathfrak{A})\big),$$

$$V(\mathfrak{A}) = \mathrm{Adm}(\overrightarrow{\mathfrak{A}}) = \mathrm{Adm}\big(N(\mathfrak{A}), E(\mathfrak{A})\big).$$

From the structurality of $\overrightarrow{\mathfrak{A}}$ it follows that $r_* \in V(\mathfrak{A})$. We have $N(\mathfrak{A}) \subseteq V(\mathfrak{A})$ and, moreover, the reverse inclusion $V(\mathfrak{A}) \subseteq N(\mathfrak{A})$ does not hold in general; we have $r_* \notin N(\mathfrak{A})$ if $\emptyset \neq E(\mathfrak{A}) \neq S$.

Products of preordered algebras

Let $\{\mathfrak{A}_t\}_{t \in T}$, where $T \neq \emptyset$, be an indexed family of similar preordered algebras ($\mathfrak{A}_t = \langle \mathscr{A}_t, \preccurlyeq_t \rangle$ for $t \in T$) such that $B_l^t(Y) \neq \emptyset$ for all $t \in T$ and all finite $Y \subseteq A_t$.
The *product* of this family is the preordered algebra

$$\mathbf{P}_{t \in T} \mathfrak{A}_t = \big\langle \mathbf{P}_{t \in T} \mathscr{A}_t, \preccurlyeq_t \big\rangle$$

with the relation $\preccurlyeq$ defined as

$$\langle x_t \rangle_{t \in T} \preccurlyeq \langle y_t \rangle_{t \in T} \quad \Leftrightarrow \quad (x_t \preccurlyeq_t y_t, \text{ for all } t \in T).$$

Theorem 2.29. $\overrightarrow{\underset{t\in T}{\mathbf{P}}\,\mathfrak{A}_t}(X) \subseteq \bigcap_{t\in T}\overrightarrow{\mathfrak{A}_t}(X),\ \ for\ every\ X \subseteq S.$

Proof. Assume that $\alpha \in \overrightarrow{\underset{t\in T}{\mathbf{P}}\,\mathfrak{A}_t}(X)$ and let $\{h_t\}_{t\in T}$ be a family of homomorphisms such that $h_t\colon \mathscr{S} \to \mathscr{A}_t$ for each $t \in T$. The mapping h defined as

$$h(\varphi) = \langle h_t(\varphi)\rangle_{t\in T}\ \ for\ \ \varphi \in S$$

is a homomorphism from $\mathscr{S}$ into $\underset{t\in T}{\mathbf{P}}\,\mathscr{A}_t$ — see Lemma 1.6. Thus, by the definition of the consequence operation $\overrightarrow{\mathfrak{A}}$, we get $h(\alpha) \in \mathbf{F}(h(X))$ where $\mathbf{F}$ denotes the filter operation in the product. Now, we apply Theorem 2.19 to get

$$h(\alpha) \in \underset{t\in T}{\mathbf{P}}\,F^t\big(\pi_t(h(X))\big)$$

where F^t stands for the filter operation in $\mathfrak{A}_t$ for each $t \in T$. Hence, we have $h_t(\alpha) = \pi_t\big(h(\alpha)\big) \in F^t\big(h_t(X)\big)$ for each $t \in T$. Since the above holds for any family $\{h_t\}_{t\in T}$ of homomorphisms, we conclude that $\alpha \in \overrightarrow{\mathfrak{A}_t}(X)$ for every $t \in T$. $\square$

It can be shown that the inclusion $\bigcap_{t\in T}\overrightarrow{\mathfrak{A}_t}(X) \subseteq \overrightarrow{\underset{t\in T}{\mathbf{P}}\,\mathfrak{A}_t}(X)$ need not be true. Let us observe, however, that this inclusion is implied by

$(\star)$ $\qquad \mathbf{F}\big(h^v(X)\big) = \underset{t\in T}{\mathbf{P}}\,F^t\big(\pi_t(h^v(X))\big),\ \ for\ every\ \ v\colon At \to \underset{t\in T}{\mathbf{P}}\,A_t.$

Indeed, assume $(\star)$ and let $\alpha \in \overrightarrow{\mathfrak{A}_t}(X)$ for each $t \in T$. Since $\pi_t \circ h^v$, for every $v\colon At \to \underset{t\in T}{\mathbf{P}}\,A_t$, is a homomorphism from $\mathscr{S}$ into $\mathscr{A}_t$, we have

$$h^v(\alpha) = \langle \pi_t \circ h^v(\alpha)\rangle_{t\in T} \in \underset{t\in T}{\mathbf{P}}\,F^t\big(\pi_t \circ h^v(X)\big)$$

and hence, by $(\star)$, we get $h^v(\alpha) \in \mathbf{F}\big(h^v(X)\big)$ for every v.

Using the above observation together with Theorem 2.19 and Corollary 2.21, we obtain

Corollary 2.30.

(i) $\overrightarrow{\underset{t\in T}{\mathbf{P}}\,\mathfrak{A}_t}(X) = \bigcap_{t\in T}\overrightarrow{\mathfrak{A}_t}(X),\ \ for\ every\ finite\ X \subseteq S;$

(ii) *If the set T is finite, then*

$$\overrightarrow{\underset{t\in T}{\mathbf{P}}\,\mathfrak{A}_t}(X) = \bigcap_{t\in T}\overrightarrow{\mathfrak{A}_t}(X),\ \ for\ every\ \ X \subseteq S.$$

Now, let us assume that $\mathfrak{A}_t = \mathfrak{A}$ for every $t \in T$. The symbol $\mathfrak{A}^T$ will stand for the product of the family $\{\mathfrak{A}_t\}_{t\in T}$, i.e., for the *$T$-power* of the preordered algebra $\mathfrak{A}$. Let us prove

Theorem 2.31. *If $Nc(S) \leqslant Nc(T)$, then $\overrightarrow{\mathfrak{A}^T}$ is a finitistic consequence operation.*

Proof. It follows from Corollary 2.30 that $\overrightarrow{\mathfrak{A}^T}(Y) = \overrightarrow{\mathfrak{A}}(Y)$ for every finite $Y \subseteq S$. Thus, in order to prove the theorem it suffices to show that

$$\overrightarrow{\mathfrak{A}^T}(X) \subseteq \bigcap\{\overrightarrow{\mathfrak{A}}(Y) : Y \in \mathrm{Fin}(X)\}, \quad \text{for each infinite } X \subseteq S.$$

Assume that $\alpha \notin \overrightarrow{\mathfrak{A}}(Y)$ for every $Y \in \mathrm{Fin}(X)$. Since $Nc\big(\mathrm{Fin}(X)\big) \leqslant Nc(T)$, there exists a mapping f from T onto $\mathrm{Fin}^*(X) = \mathrm{Fin}(X) \setminus \{\emptyset\}$. In the sequel we will write X_t instead of $f(t)$.

It has been assumed that $\alpha \notin \overrightarrow{\mathfrak{A}}(X_t)$ for every $t \in T$. Hence, we can choose (by the Axiom of Choice) for each $t \in T$ a homomorphism $h_t \colon \mathscr{S} \to \mathscr{A}$ such that

$$(\star) \qquad\qquad\qquad h_t(\alpha) \notin F\big(h_t(X)\big).$$

Let h be the product of the homomorphisms $\{h_t\}_{t \in T}$ — see Lemma 1.6 — i.e., let $h \colon \mathscr{S} \to \mathscr{A}^T$ and

$$h(\varphi) = \langle h_t(\varphi) \rangle_{t \in T}, \quad \text{for every } \varphi \in S.$$

It is easy to observe that for every $t \in T$,

$$h(X_t) \subseteq \mathop{\mathbf{P}}_{s \in T} \pi_s\big(h(X_t)\big) = \mathop{\mathbf{P}}_{s \in T} \pi_s \circ h(X_t) = \mathop{\mathbf{P}}_{s \in T} h_s(X_t) \subseteq \mathop{\mathbf{P}}_{s \in T} F\big(h_s(X_t)\big)$$

and $h(\alpha) \notin \mathop{\mathbf{P}}_{s \in T} F\big(h_s(X_t)\big)$ by $(\star)$.

But $\mathop{\mathbf{P}}_{s \in T} F\big(h_s(X_t)\big)$ is a filter in $\mathfrak{A}^T$ by Theorem 2.18 (i). Thus, $h(\alpha) \notin \mathbf{F}\big(h(X_t)\big)$ for every $t \in T$ where $\mathbf{F}$ is the filter operation in $\mathfrak{A}^T$. Since $\{X_t : t \in T\} = \mathrm{Fin}^*(X)$, we conclude that $h(\alpha) \notin \bigcup\{\mathbf{F}\big(h(Y)\big) : Y \in \mathrm{Fin}^*(X)\}$ and hence $h(\alpha) \notin \mathbf{F}\big(h(X)\big)$ on the basis of Corollary 2.12 (iv). $\qquad\square$

The above theorem suggests that some properties of the consequence operation determined by the product of preordered algebras may depend upon the number of components in the product.

2.3 Logical matrices

This section deals with logical matrices which will be defined as abstract algebras with distinguished elements. As it will appear, matrices can also be understood as a special case of preordered algebras. Despite these connections we will preserve the traditional terminology referring to the work of J. Łukasiewicz and A. Tarski [66], 1930 and their followers.

Matrix consequences

A *logical matrix* is a pair $\mathfrak{M} = \langle \mathscr{A}, A^* \rangle$, where $\mathscr{A}$ is an abstract algebra and A^* is a subset of the *universe* of $\mathscr{A}$, i.e., $A^* \subseteq A$. Any $a \in A^*$ is called a *distinguished element* of the matrix $\mathfrak{M}$.

The set of $\mathfrak{M}$-*valid formulas* ($\mathfrak{M}$-tautologies) is defined in the standard way:

$$\alpha \in E(\mathfrak{M}) \quad \Leftrightarrow \quad \left(h^v(\alpha) \in A^*, \ \text{for every} \ v\colon At \to A \right).$$

The following definition of the *matrix consequence* $\overrightarrow{\mathfrak{M}}$ is due to J. Łoś and R. Suszko [64], 1958.

Definition 2.32. For every $X \subseteq S$ and every $\alpha \in S$,

$$\alpha \in \overrightarrow{\mathfrak{M}}(X) \quad \Leftrightarrow \quad \left(h^v(X) \subseteq A^* \Rightarrow h^v(\alpha) \in A^*, \ \text{for every} \ v\colon At \to A \right).$$

Observe that $E(\mathfrak{M}) = \overrightarrow{\mathfrak{M}}(\emptyset)$. We shall compare the just defined notion of the matrix consequence with the notion of the consequence operation generated by a preordered algebra.

Consider any abstract algebra $\mathscr{A}$ with the universe A. Let us recall that we deal only with algebras similar to the algebra $\mathscr{S}$ of a fixed propositional language. Distinguishing a set $A^* \subseteq A$ we get the matrix $\mathfrak{M} = \langle \mathscr{A}, A^* \rangle$. First, let us consider the case when A^* is empty or when $A^* = A$. It is easy to notice that the matrix $\langle \mathscr{A}, A \rangle$ generates the inconsistent consequence operation, i.e., $\overrightarrow{\mathfrak{M}}(X) = S$ for every $X \subseteq S$ and the matrix-consequence of $\mathfrak{M} = \langle \mathscr{A}, \emptyset \rangle$ is

$$\overrightarrow{\mathfrak{M}}(X) = \begin{cases} \emptyset & \text{if} \ X = \emptyset \\ S & \text{if} \ X \neq \emptyset. \end{cases}$$

Both operations are trivial examples of matrix-consequences.

Let us assume that $\emptyset \subsetneq A^* \subsetneq A$ and define a preorder relation $\preccurlyeq$ on A by

$$x \preccurlyeq y \quad \Leftrightarrow \quad x \notin A^* \vee y \in A^*, \qquad \text{for} \ x, y \in A.$$

Let us prove that A^* is the only proper filter in the preordered set $\langle A, \preccurlyeq \rangle$. First observe that

(i) $\text{Great}(A) = A^*$,

(ii) $\text{Least}(A) = A \setminus A^*$.

Therefore, by Lemma 2.13 (i) and (iii), we conclude that

$$F(X) = \begin{cases} A^* & \text{if} \ X \subseteq A^* \\ A & \text{otherwise} \end{cases}$$

and hence A^* is the only proper filter in $\langle A, \preccurlyeq \rangle$.

Thus, by Lemma 2.23 and Definition 2.32 we get $\overrightarrow{\mathfrak{M}}(X) = \overrightarrow{\mathfrak{A}}(X)$, where $\mathfrak{A} = \langle \mathscr{A}, \preccurlyeq \rangle$ is the preordered algebra generated by the set A^*. This is the very connection between preordered algebras and matrices. It means, in particular, that all notions and results concerning preordered algebras can also be applied to logical matrices. For example, it follows from Corollary 2.24 that $\overrightarrow{\mathfrak{M}}$ is a structural consequence operation.

Moreover, analogously as for algebras, one may define *rules normal* in a matrix $\mathfrak{M}$ and *rules valid* in $\mathfrak{M}$:

$$r \in N(\mathfrak{M}) \quad \Leftrightarrow \left(\alpha \in \overrightarrow{\mathfrak{M}}(\Pi), \text{ for each } \langle \Pi, \alpha \rangle \in r \right);$$

$$r \in V(\mathfrak{M}) \quad \Leftrightarrow \left(\Pi \subseteq E(\mathfrak{M}) \Rightarrow \alpha \in E(\mathfrak{M}), \text{ for each } \langle \Pi, \alpha \rangle \in r \right).$$

We quote without proof the following simple lemma:

Lemma 2.33.

(i) $N(\mathfrak{M}) = \mathrm{Der}(\overrightarrow{\mathfrak{M}})$;

(ii) $V(\mathfrak{M}) = \mathrm{Adm}(\overrightarrow{\mathfrak{M}})$;

(iii) $r_* \in V(\mathfrak{M}) \setminus N(\mathfrak{M}) \quad if \quad \emptyset \subsetneq A^* \subsetneq A$.

Finite matrices

Let $\mathfrak{M} = \langle \mathscr{A}, A^* \rangle$ be a logical matrix and let $A^* \neq \emptyset$. It has been proved above that there exists a preorder $\preccurlyeq$ on S such that $\overrightarrow{\mathfrak{M}} = \overrightarrow{\mathfrak{A}}$, where $\mathfrak{A} = \langle \mathscr{A}, \preccurlyeq \rangle$. Hence, as an immediate corollary of Theorem 2.27 we obtain the following result due to J. Łoś and R. Suszko [64], 1958

Corollary 2.34. *If $\mathfrak{M}$ is a finite matrix, then $\overrightarrow{\mathfrak{M}}$ is a finitistic consequence operation.*

Let us introduce the following definition:

Definition 2.35. A set $X \subseteq S$ is said to be satisfiable in $\mathfrak{M} = \langle \mathscr{A}, A^* \rangle$, in symbols $X \in \mathrm{Sat}(\mathfrak{M})$, if and only if there is a valuation $v \colon At \to A$ such that

$$h^v(X) \subseteq A^*.$$

We prove now the following theorem.

Theorem 2.36. *If $\mathfrak{M}$ is a finite matrix and $X \subseteq S$, then*

$$X \in \mathrm{Sat}(\mathfrak{M}) \quad \Leftrightarrow \quad \left(Y \in \mathrm{Sat}(\mathfrak{M}), \quad \text{for each } Y \in \mathrm{Fin}(X) \right).$$

Proof. The implication ($\Rightarrow$) is obvious. Assume now that $\mathfrak{M} = \langle \mathscr{A}, A^* \rangle$ and that $Y \in \mathrm{Sat}(\mathfrak{M})$ for each $Y \in \mathrm{Fin}(X)$. Let V be the family of all valuations $v \colon At \to A$ and define

(i) $V(Y) = \{ v \in V : h^v(Y) \subseteq A^* \}$.

It is obvious that

(ii) $Y \subseteq Y_0 \implies V(Y_0) \subseteq V(Y)$.

It follows from the assumption that $V(Y)$ is non-empty for every $Y \in \mathrm{Fin}(X)$. Hence, by (ii),

$$\emptyset \neq V(Y_1 \cup \ldots \cup Y_k) \subseteq V(Y_1) \cap \ldots \cap V(Y_k)$$

for every $Y_1, \ldots, Y_k \in \mathrm{Fin}(X)$. Thus, the family $\{V(Y) : Y \in \mathrm{Fin}(X)\}$ has the finite intersection property and then there is a maximal filter H in 2^V containing all sets $V(Y)$ for $Y \in \mathrm{Fin}(X)$ — see Theorem 1.18.

Since V is the greatest element in 2^V, we have $V \in H$ and hence

$$\bigcup_{a \in A} \{v \in V : v(p) = a\} = V \in H$$

for every variable $p \in At$. But H is a prime filter as an ultrafilter (see Corollary 1.22) and A is finite; therefore

(iii) for every $p \in At$ there is an element $a \in A$ such that

$$\{v \in V : v(p) = a\} \in H.$$

Moreover,

(iv) $\{v \in V : v(p) = a\} \cap \{v \in V : v(p) = b\} = \emptyset \notin H$ if $a \neq b$.

We define a valuation $w \colon At \to A$ by taking

$$w(p) = a \quad \Leftrightarrow \quad \{v \in V : v(p) = a\} \in H.$$

From (iv), (iii) it follows that the valuation w is well defined and what is more,

(v) $\{v \in V : v(p) = w(p)\} \in H$ for every $p \in At$.

Let us consider $Y \in \mathrm{Fin}(X)$. Since $At(Y)$ is finite and since H is a filter, we obtain by (v),

$$\{v \in V : v|_{At(Y)} = w_{At(Y)}\} = \bigcap_{p \in At(Y)} \{v \in V : v(p) = w(p)\} \in H.$$

Thus,

(vi) $\{v \in V : v|_{At(Y)} = w|_{At(Y)}\} \cap V(Y) \in H$.

By (vi) and (i) there is a valuation $v \colon At \to A$ such that $v|_{At(Y)} = w|_{At(Y)}$ and $h^v(Y) \subseteq A^*$. Hence, $h^w(Y) \subseteq A^*$ for each finite set $Y \subseteq X$, which yields $h^w(X) \subseteq A^*$. $\qquad\square$

The method of proving Theorem 2.36 (and also Theorem 2.27) can be generalized. If one wants to prove that $\overrightarrow{\mathfrak{M}}$ is finitistic, it is enough to verify that for every ultrafilter H in 2^V the condition (iii) from the proof of 2.36 is fulfilled.

Constructions of matrices

Let $\mathfrak{M} = \langle \mathscr{A}, A^* \rangle$ and $\mathfrak{N} = \langle \mathscr{B}, B^* \rangle$ be two similar matrices. $\mathfrak{N}$ is called a *submatrix* of $\mathfrak{M}$ ($\mathfrak{N} \subseteq \mathfrak{M}$) if and only if $\mathscr{B}$ is a subalgebra of $\mathscr{A}$ and $B^* = B \cap A^*$.

The matrices $\mathfrak{M}$, $\mathfrak{N}$ are said to be *isomorphic* ($\mathfrak{M} \cong \mathfrak{N}$) iff there is an isomorphism h from $\mathscr{A}$ onto $\mathscr{B}$ such that

$$x \in A^* \quad \Leftrightarrow \quad h(x) \in B^*, \text{ for each } x \in A.$$

It is a trivial fact that

Corollary 2.37.

(i) $\mathfrak{N} \subseteq \mathfrak{M} \quad \Rightarrow \quad \overrightarrow{\mathfrak{M}} \leqslant \overrightarrow{\mathfrak{N}}$;

(ii) $\mathfrak{N} \cong \mathfrak{M} \quad \Rightarrow \quad \overrightarrow{\mathfrak{M}} = \overrightarrow{\mathfrak{N}}$.

In the next theorem all $\overrightarrow{\mathfrak{M}} \circ Sb$ closed sets are characterized by means of submatrices of $\mathfrak{M}$ (see [131], 1978).

Theorem 2.38. *For every* $X \subseteq S$,

$$\overrightarrow{\mathfrak{M}}(Sb(X)) = \bigcap \{E(\mathfrak{N}) : \mathfrak{N} \subseteq \mathfrak{M} \wedge X \subseteq E(\mathfrak{N})\}.$$

Proof. The inclusion ($\subseteq$) follows immediately from Corollary 2.37 (i). To prove ($\supseteq$) assume that $\alpha \notin \overrightarrow{\mathfrak{M}}(Sb(X))$ and let $\mathfrak{M} = \langle \mathscr{A}, A^* \rangle$. We have to show that $X \subseteq E(\mathfrak{N})$ and $\alpha \notin E(\mathfrak{N})$ for some submatrix $\mathfrak{N}$ of $\mathfrak{M}$.

If $X = \emptyset$, then $X \subseteq E(\mathfrak{M})$ and $\alpha \notin \overrightarrow{\mathfrak{M}}(Sb(\emptyset)) = E(\mathfrak{M})$. Now let $X \neq \emptyset$. There exists a homomorphism h from $\mathscr{S}$ into $\mathscr{A}$ such that $h(Sb(X)) \subseteq A^*$ and $h(\alpha) \notin A^*$. Let $\mathscr{B}$ be the subalgebra of $\mathscr{A}$ generated by the set $h(S)$. It follows from Lemma 1.2 that $h(S)$ is the universe of $\mathscr{B}$ since $h(S)$ is closed under the operation of $\mathscr{A}$.

We define $\mathfrak{N} = \langle \mathscr{B}, h(S) \cap A^* \rangle$; clearly $\mathfrak{N} \subseteq \mathfrak{M}$. Let us prove that $X \subseteq E(\mathfrak{N})$. Suppose that $v : At \to h(S)$. Using the axiom of choice we can choose a substitution $e : At \to S$ such that

$$e(\gamma) \in h^{-1}(\{v(\gamma)\}), \quad \text{for every } \gamma \in At.$$

Thus $h^v = h \circ h^e$ and hence $h^v(X) = h(h^e(X)) \subseteq h(Sb(X)) \subseteq A^*$. This proves that $X \subseteq E(\mathfrak{N})$. Moreover, $h(\alpha) \notin A^*$ and $h : S \to h(S)$, hence $\alpha \notin E(\mathfrak{N})$. $\qquad \square$

A binary relation R on A is called a *congruence* on the matrix $\mathfrak{M} = \langle \mathscr{A}, A^* \rangle$ if and only if R is a congruence on the algebra $\mathscr{A}$ satisfying the condition

$$xRy \wedge x \in A^* \quad \Rightarrow \quad y \in A^*.$$

Every congruence R on $\mathfrak{M}$ determines the quotient matrix

$$\mathfrak{M}/R = \langle \mathscr{A}/R, A^*/R \rangle$$

where $\mathscr{A}/R$ is the quotient algebra and $A^*/R = \{[a]_R : a \in A^*\}$. The next theorem can be found in [142], 1973.

Theorem 2.39. *If R is a congruence of the matrix $\mathfrak{M}$, then*

$$\overrightarrow{\mathfrak{M}} = \overrightarrow{\mathfrak{M}/R}.$$

Proof. $(\leqslant)$: Suppose that $\alpha \notin \overrightarrow{\mathfrak{M}/R}(X)$. Then there exists a homomorphism h from $\mathscr{S}$ into $\mathscr{A}/R$ such that $h(X) \subseteq A^*/R$ and $h(\alpha) \notin A^*/R$. By the axiom of choice, there is a mapping $v \colon At \to A$ such that $v(\gamma) \in h(\gamma)$ for each variable $\gamma \in At$. Let us extend the mapping v to the homomorphism $h^v \colon \mathscr{S} \to \mathscr{A}$. Then $h^v(\varphi) \in h(\varphi)$ for every $\varphi \in S$ and hence $h^v(X) \subseteq A^*$ and $h^v(\alpha) \notin A^*$.

$(\geqslant)$: Assume that $\alpha \notin \overrightarrow{\mathfrak{M}}(X)$. Then $h^v(X) \subseteq A^*$ and $h^v(\alpha) \notin A^*$ for some $v \colon At \to A$. Let $h_R \colon A \to A/R$ be the canonical homomorphism from $\mathscr{A}$ onto $\mathscr{A}/R$, i.e., $h_R(a) = [a]_R$ for $a \in A$. The composition $h = h_R \circ h^v$ is a homomorphism from $\mathscr{S}$ into $\mathscr{A}/R$ and, moreover, we have $h(X) = h_R(h^v(X)) \subseteq A^*/R$ and $h(\alpha) = [h^v(\alpha)]_R \notin A^*/R$. Thus $\alpha \notin \overrightarrow{\mathfrak{M}/R}(X)$. $\qquad\square$

Let $\{\mathfrak{M}_t\}_{t \in T}$, where $T \neq \emptyset$, be an indexed family of similar matrices

$$\mathfrak{M}_t = \langle \mathscr{A}_t, A_t^* \rangle.$$

The matrix

$$\mathbf{P}_{t \in T} \mathfrak{M}_t = \langle \mathbf{P}_{t \in T} \mathscr{A}_t, \mathbf{P}_{t \in T} A_t^* \rangle$$

is called the *product of the family* $\{\mathfrak{M}_t\}_{t \in T}$.

It will be additionally assumed that A_t^* is non-empty for every $t \in T$. Let $\preccurlyeq_t$ be the preorder on A_t (the universe of $\mathscr{A}_t$) determined by the set A_t^*, i.e.,

$$x \preccurlyeq_t y \quad \Leftrightarrow \quad x \notin A_t^* \lor y \in A_t^*, \qquad \text{for } x, y \in A_t.$$

It has been proved that A_t^* is the only proper (if $A_t^* \neq A_t$) filter in $\langle \mathscr{A}_t, \preccurlyeq_t \rangle$. The product $\mathbf{P}_{t \in T}\langle \mathscr{A}_t, \preccurlyeq_t \rangle$ may contain, however, more than one filter.

The product $\mathbf{P}_{t \in T}\langle \mathscr{A}_t, A^* \rangle$ is a logical matrix and $\mathbf{P}_{t \in T}\langle \mathscr{A}_t, \preccurlyeq_t \rangle$ is a preordered algebra. Some connections between both structures, clearly, occur. Namely, observe that $\mathbf{P}_{t \in T} A_t^*$ is the least filter in $\mathbf{P}_{t \in T}\langle \mathscr{A}_t, \preccurlyeq_t \rangle$. In other words,

$(\star)$ $$\mathbf{P}_{t \in T} A_t^* = \mathbf{F}(\emptyset)$$

where $\mathbf{F}$ is the filter operation in the product. Thus, as an immediate result of Corollary 2.30 (i), we obtain (see Jaśkowski [44], 1936)

Corollary 2.40. $\quad E(\mathbf{P}_{t \in T} \mathfrak{M}_t) = \bigcap\{E(\mathfrak{M}_t) : t \in T\}.$

Proof. By 2.30 (i), $\overrightarrow{\mathbf{P}\,\mathfrak{A}_t}(\emptyset) = \bigcap\{\overrightarrow{\mathfrak{A}_t}(\emptyset) : t \in T\}$ where $\mathfrak{A}_t = \langle \mathscr{A}_t, \preccurlyeq_t \rangle$. But $\overrightarrow{\mathfrak{M}_t} = \overrightarrow{\mathfrak{A}_t}$ and hence $E(\mathfrak{M}_t) = \overrightarrow{\mathfrak{A}_t}(\emptyset)$ for every $t \in T$. On the other hand, according to Definition 2.22, $\alpha \in \overrightarrow{\mathbf{P}\,\mathfrak{A}_t}(\emptyset)$ iff $h^v(\alpha) \in \mathbf{F}\big(h^v(\emptyset)\big)$ for every $v \colon At \to \underset{t \in T}{\mathbf{P}}\,A_t$. Thus, by $(\star)$, $\overrightarrow{\mathbf{P}\,\mathfrak{A}_t}(\emptyset) = E(\underset{t \in T}{\mathbf{P}}\,\mathfrak{M}_t)$, which completes the proof. $\qquad\square$

The above theorem can be strengthened to the following one, see [67], 1973.

Theorem 2.41. *For every* $X \subseteq S$,

$$
\overrightarrow{\underset{t \in T}{\mathbf{P}}\,\mathfrak{M}_t}(X) = \begin{cases} \bigcap\{\overrightarrow{\mathfrak{M}_t}(X) : t \in T\} & \text{if } X \in \mathrm{Sat}(\mathfrak{M}_t) \text{ for each } t \in T \\ S & \text{otherwise.} \end{cases}
$$

Proof. ($\supseteq$): Let us assume that $\alpha \notin \overrightarrow{\underset{t \in T}{\mathbf{P}}\,\mathfrak{M}_t}(X)$. Then there exists a homomorphism h from $\mathscr{S}$ into the product such that

$$
h(X) \subseteq \underset{t \in T}{\mathbf{P}}\,A_t^* \quad \text{and} \quad h(\alpha) \notin \underset{t \in T}{\mathbf{P}}\,A_t^*.
$$

Hence $\pi_t\big(h(X)\big) \subseteq A_t^*$ for each $t \in T$ and $\pi_s\big(h(\alpha)\big) \notin A_s^*$ for some $s \in T$. But the mapping $\pi_t \circ h$, for $t \in T$, is a homomorphism from $\mathscr{S}$ into $\mathscr{A}_t$, so $X \in \mathrm{Sat}(\mathfrak{M}_t)$ for each $t \in T$ and $\alpha \notin \overrightarrow{\mathfrak{M}_s}(X)$ for some $s \in T$. Thus $\alpha \notin \bigcap_{t \in T} \overrightarrow{\mathfrak{M}_t}(X)$.

($\subseteq$): Assume that $X \in \mathrm{Sat}(\mathfrak{M}_t)$ for each $t \in T$ and let $\alpha \notin \bigcap_{t \in T} \overrightarrow{\mathfrak{M}_t}(X)$. From these assumptions it follows that there exists a family $\{h_t\}_{t \in T}$ of homomorphisms such that $h_t \colon \mathscr{S} \to \mathscr{A}_t$, $h_t(X) \subseteq A_t^*$ for each $t \in T$ and $h_s(\alpha) \notin A_s^*$ for some $s \in T$. Then the mapping h defined by

$$
h(\varphi) = \langle h_t(\varphi) \rangle_{t \in T} \quad \text{for } \varphi \in S
$$

is a homomorphism from $\mathscr{S}$ into $\underset{t \in T}{\mathbf{P}}\,\mathscr{A}_t$ (see Lemma 1.6) and, clearly, $h(X) \subseteq \underset{t \in T}{\mathbf{P}}\,A_t^*$ and $h(\alpha) \notin \underset{t \in T}{\mathbf{P}}\,A_t^*$. This means that $\alpha \notin \overrightarrow{\underset{t \in T}{\mathbf{P}}\,\mathfrak{M}_t}(X)$. $\qquad\square$

Lindenbaum matrices

Let $\langle R, X \rangle$ be a system of propositional logic, i.e., let $R \subseteq \mathscr{R}_S$ and $X \subseteq S$. Then

$$
\mathfrak{M}^{R,X} = \langle \mathscr{S}, Cn(R, X) \rangle
$$

is called the *Lindenbaum matrix* of the system $\langle R, X \rangle$.

Lemma 2.42. *For every* $X \subseteq S$ *and* $R \subseteq \mathscr{R}_S$,

$$
E(\mathfrak{M}^{R,X}) = \{\alpha : Sb(\alpha) \subseteq Cn(R, X)\}.
$$

This lemma follows immediately from the involved definitions.

Corollary 2.43. *If $r_* \in \mathrm{Adm}(R, X)$, then $E(\mathfrak{M}^{R,X}) = Cn(R, X)$.*

The next lemma states some connections between rules normal and valid in the Lindenbaum matrix.

Lemma 2.44. *If $r_* \in \mathrm{Adm}(R, X)$, then*

$$V(\mathfrak{M}^{R,X}) \cap \mathrm{Struct} \subseteq N(\mathfrak{M}^{R,X}).$$

Proof. Assume that r is a structural rule and that $r \in V(\mathfrak{M}^{R,X})$. Let $\langle \Pi, \alpha \rangle \in r$ and suppose that $h^e(\Pi) \subseteq Cn(R, X)$ for some $e \colon At \to S$.

We need to show that $h^e(\alpha) \in Cn(R, X)$.

By Corollary 2.43, $h^e(\Pi) \subseteq E(\mathfrak{M}^{R,X})$. Moreover, $\langle h^e(\Pi), h^e(\alpha) \rangle \in r$ since r is structural. Then, by the definition of $V(\mathfrak{M})$ and by 2.43, we obtain $h^e(\alpha) \in E(\mathfrak{M}^{R,X}) = = Cn(R, X)$. $\square$

The just defined Lindenbaum matrix $\mathfrak{M}^{R,X}$, though related to, should not be mixed with the so-called Lindenbaum–Tarski algebra, defined below, which usually appears in proofs of completeness theorems for various propositional logics. First of all, the Lindenbaum matrix is defined for arbitrary logical system whereas the construction of the Lindanbaum–Tarski algebra requires certain assumptions.

Suppose that $\langle R, X \rangle$ is a propositional logic formalized over the standard language $\mathscr{S}_2 = \langle S_2, \to, +, \cdot, \sim \rangle$ (or any of its implicative sublanguage). We define a binary relation on S_2 by

$$\alpha \sim_{R,X} \beta \quad \text{iff} \quad \alpha \to \beta \ , \ \beta \to \alpha \in Cn(R, X)$$

and we will write $\sim$ (or $\sim_X$) instead of $\sim_{R,X}$ if there is no danger of confusion. If we assume that the rule r_0 and the following formulas are derivable in $\langle R, X \rangle$,

$$(p \to q) \to ((q \to s) \to (p \to s))$$
$$(q \to s) \to ((p \to q) \to (p \to s))$$
$$p \to p$$
$$p \cdot q \to p$$
$$p \cdot q \to q$$
$$(p \to q) \to ((p \to s) \to (p \to q \cdot s))$$
$$p \to p + q$$
$$q \to p + q$$
$$(p \to s) \to ((q \to s) \to (p + q \to s))$$
$$(p \to q) \to (\sim q \to \sim p),$$

then the relation $\sim_{R,X}$ turns out to be a congruence relation on the Lindenbaum matrix $\mathfrak{M}^{R,X}$ (we leave this known theorem without proof). Then the quotient algebra

$$\mathscr{S}_2/\!\sim = \langle S_2/\!\sim, \dot{\to}, \cup, \cap, - \rangle$$

is called the *Lindenbaum–Tarski algebra* of the system $\langle R, X \rangle$ where

$$[\alpha] \cup [\beta] = [\alpha + \beta]$$
$$[\alpha] \cap [\beta] = [\alpha \cdot \beta]$$
$$[\alpha] \dot{\to} [\beta] = [\alpha \to \beta]$$
$$-[\alpha] = [\sim \alpha].$$

Next, it can also be proved that $\langle S/\!\!\sim, \cup, \cap \rangle$ is a lattice with the lattice ordering $\leqslant$ defined as

$$[\alpha] \leqslant [\beta] \quad \Leftrightarrow \quad (\alpha \to \beta) \in Cn(R, X).$$

If we assume additionally that $q \to (p \to p) \in Cn(R, X)$, then the unit element exists in the Lindenbaum–Tarski algebra. Moreover, if $p \to (q \to p) \in Cn(R, X)$, then

$$[\alpha] = 1 \quad \Leftrightarrow \quad \alpha \in Cn(R, X).$$

The existence of the zero element will be guaranteed by $\sim (p \to p) \to q \in Cn(R, X)$. From the assumptions that $\sim (p \to p) \to q$, $p \to (q \to p)$, $(p \to\sim q) \to (q \to\sim p)$ are in $Cn(R, X)$ it will follow that

$$[\alpha] = 0 \quad \Leftrightarrow \quad \sim \alpha \in Cn(R, X).$$

The above concerns, as well, all systems defined in implicative sublanguages of $\mathscr{S}_2$. For instance, it concerns $\mathscr{S}_1$ if we reject from axioms the formula containing negation and remove the above fragment about the existence of the zero-element.

2.4 Adequacy

The present section is concerned with connections between a propositional logic and its semantics. Our interest in algebraic structures such as preordered algebras, matrices, lattices, ... etc. stems from some problems which arise in the theory of formal systems. The notion of abstract algebra has proved to be a useful tool in the investigations of non-classical logics. The algebraic approach to logic originated in the efforts of logicians to find some new connections between algebra and logic and to enrich the proof-theoretical tools of logic. Adequacy consists in showing a parallelism between some algebraic structures on the one hand, and a certain logic on the other. For this purpose, however, different kinds of structures have been used. In the sequel we will discuss some of them. A special attention will be given to the matrix semantics of the most important propositional logics.

Adequacy of matrices

Let $\langle R, X \rangle$ be a system of a propositional logic formalized in a fixed propositional language and let $\mathfrak{M}$ be a logical matrix similar to the algebra of the language. *Completeness* of $\langle R, X \rangle$ with respect to the matrix $\mathfrak{M}$ ($\mathfrak{M}$-completeness) is understood

as the equality
$$E(\mathfrak{M}) = Cn(R, X) = Cn_{RX}(\emptyset).$$

If the system $\langle R, X \rangle$ fulfills this condition, then the matrix $\mathfrak{M}$ is said to be *(weakly) adequate* for $\langle R, X \rangle$.

Since $E(\mathfrak{M})$ is closed with respect to the substitution rule r_* (see Lemma 2.33), the problem of $\mathfrak{M}$-adequacy of a logic can be put forward only if $r_* \in$ Adm(R, X). A general solution of this problem is given by the well-known Lindenbaum theorem which states that $Cn(R, X) = E(\mathfrak{M}^{R,X})$ for every propositional logic $\langle R, X \rangle$ such that $r_* \in$ Adm(R, X) — see Corollary 2.43. We also get

Corollary 2.45. *For each propositional logic $\langle R, X \rangle$ with $r_* \in$ Adm(R, X) and $R \setminus \{r_*\} \subseteq$ Struct there is a matrix $\mathfrak{M}$ such that:*

(i) $Cn(R, X) = E(\mathfrak{M})$;

(ii) $R \setminus \{r_*\} \subseteq N(\mathfrak{M})$.

The above Lindenbaum theorem seems to prove that the whole question is quite trivial. However, it is not so since logicians are not looking for adequate matrices of any kind but for adequate matrices which fulfill some special conditions. For example, one may postulate that the universe of the adequate matrix ought to be built up from a special kind of objects, such as, e.g., subsets of some topological space, sequences of some natural numbers, ... etc. One may also look for a finite adequate matrix $\mathfrak{M}$ (if such exists) and what is more one may render $\mathfrak{M}$ to be a minimal matrix, i.e.,

$$E(\mathfrak{M}) = E(\mathfrak{N}) \quad \Rightarrow \quad Nc(\mathfrak{M}) \leqslant Nc(\mathfrak{N}).$$

As an illustration let us mention the classical logic $\langle R_{0*}, A_2 \rangle$ and the well-known theorem by Post which says that $\langle R_{0*}, A_2 \rangle$ is complete with respect to the ordinary two-element matrix

$$\mathfrak{M}_2 = \langle \langle \{0, 1\}, f_2^{\rightarrow}, f_2^{+}, f_2^{\cdot}, f_2^{\sim} \rangle, \{1\} \rangle.$$

$\mathfrak{M}_2$ is, of course, a minimal adequate matrix for classical logic. Note that the above characterization of classical logic gives a method of deciding whether a given formula α is (or is not) derivable on the grounds of $\langle R_{0*}, A_2 \rangle$.

In the case of the modal system $S5$, an adequate matrix was defined by M. Wajsberg. Let $\mathfrak{M}_{S5} = \langle \langle \{0, 1\}^{\mathbb{N}}, g^{\rightarrow}, g^{+}, g^{\cdot}, g^{\sim} \rangle, \{1\} \rangle$ where $\mathbb{N}$ is the set of natural numbers and

$$\mathbf{1} = \langle x_n \rangle_{n \in \mathbb{N}} \Leftrightarrow x_n = 1 \text{ for all } n \in \mathbb{N}$$

$$g^{\cdot}(\langle x_n \rangle_{n \in \mathbb{N}}, \langle y_n \rangle_{n \in \mathbb{N}}) = \langle \min\{x_n, y_n\} \rangle_{n \in \mathbb{N}}$$

$$g^{+}(\langle x_n \rangle_{n \in \mathbb{N}}, \langle y_n \rangle_{n \in \mathbb{N}}) = \langle \max\{x_n, y_n\} \rangle_{n \in \mathbb{N}}$$

$$g^{\sim}(\langle x_n \rangle_{n \in \mathbb{N}}) = \langle 1 - x_n \rangle_{n \in \mathbb{N}}$$

$$g^{\rightarrow}(\langle x_n \rangle_{n \in \mathbb{N}}, \langle y_n \rangle_{n \in \mathbb{N}}) = \begin{cases} \mathbf{1} & \text{if } x_n \leqslant y_n \text{ for all } n \in \mathbb{N} \\ g^{\sim}(\mathbf{1}) & \text{otherwise.} \end{cases}$$

It should be observed that $g^+, g^\cdot, g^\sim$ are the operations of the product of matrices $\mathfrak{M}_2$ (more specifically, the operations of the power matrix $(\mathfrak{M}_2)^{\mathbb{N}}$).

The just defined matrix $\mathfrak{M}_{S5}$ is adequate for the modal system $S5$, i.e., $E(\mathfrak{M}_{S5}) = Cn(R_{0a*}, A_{S5})$. Let us add that $\langle R_{0a*}, A_{S5} \rangle$ has no finite adequate matrix.

On the other hand, another approach to the whole problem is also possible. A propositional logic can be defined with the help of a logical matrix $\mathfrak{M}$ and we may be looking for a $\mathfrak{M}$-complete system $\langle R, X \rangle$. Of course, the pair $\langle N(\mathfrak{M}), E(\mathfrak{M}) \rangle$ is $\mathfrak{M}$-complete. Our aim, however, is not to reduce the notion of $\mathfrak{M}$-validity to syntactic framework but to get a characterization of $E(\mathfrak{M})$. For example, we may be looking for a finite set X and for a finite set R of standard rules such that $Cn(R \cup \{r_*\}, X) = E(\mathfrak{M})$.

The work of M. Wajsberg [129], 1935, contains a result concerning finite axiomatizability (in the above sense) of a broad class of finite matrices. The proof of this theorem contains an algorithm for searching of a finite set of axioms for any given finite matrix. But the algorithm is very difficult and practically useless. Let us remark that it is not true that for every logical matrix there exist a finite set X of formulas and a finite set R of standard rules such that $Cn(R \cup \{r_*\}, X) = E(\mathfrak{M})$, even if we assume that $\mathfrak{M}$ is finite. The best counterexample is due to K. Pałasińska [72], 1974. She showed a 3-element matrix, in the pure implicational language $\{\rightarrow\}$, with one distinguished element which is not finitely axiomatizable in any reasonable way. The matrix is given by the following table

$\rightarrow$	0	1	2
0	1(2)	2	2
1	2	2	2
2	1	2	2

where 2 is distinguished and 1(2) in the first row means that one can put either 1 or 2 there. So, there are given in fact two 3-element matrices with the required property. The counterexample is optimal as it follows from certain results in W. Rautenberg [108], 1981 that all propositional logics characterized by two-element matrices enjoy the finite axiomatizabilty property.

The class of matrices considered in Wajsberg's paper [129], 1935 contains all finite *Łukasiewicz matrices*, i.e., all systems

$$\mathfrak{M}_n = \langle \langle A_n, f^\rightarrow, f^+, f^\cdot, f^\sim \rangle, \{1\} \rangle \quad \text{for} \quad n \in \mathbb{N}$$

where $A_n = \left\{ \dfrac{k}{n-1} : k = 0, 1, \ldots, n-1 \right\}$ and

$$f^\rightarrow(x, y) = \min\{1 - x + y, 1\}$$
$$f^+(x, y) = \max\{x, y\}$$
$$f^\cdot(x, y) = \min\{x, y\}$$
$$f^\sim(x) = 1 - x.$$

It should be observed that $\mathfrak{M}_2$ is the classical matrix. In the first chapter we provided the n-valued Łukasiewicz logic with a set of axioms, i.e.,

$$Cn(R_{0*}, \mathbb{L}_n) = E(\mathfrak{M}_n).$$

For the proof of the above equality see [128], 1988. The n-valued Łukasiewicz matrix $\mathfrak{M}_n$ is thus adequate for the n-valued Łukasiewicz logic $\langle R_{0*}, \mathbb{L}_n \rangle$; it is in fact a minimal adequate matrix for this logic.

Wajsberg also confirmed Łukasiewicz's conjecture and showed the completeness theorem for the infinite valued logic

$$E(\mathfrak{M}_\infty) = Cn(R_{0*}, \mathbb{L}_\infty)$$

where the infinite valued matrix $\mathfrak{M}_\infty$ is defined — just as the finite ones — on the real interval $[0, 1]$. (NB. In Łukasiewicz's original conjecture the set $\mathbb{L}_\infty$ was extended with the formula $((p \to q) \to (q \to p)) \to (q \to p)$, which was shown to be dependent by D. Meredith.) Unfortunately, Wajsberg's proof has never been published and the first (non-elementary) proof of this theorem was given by C.C. Chang [10], 1955. An elementary and easy proof of this theorem can be found in R. Cignoli, D. Mundici [11], 1997.

In the opinion of many logicians in the first half of the 20th century the problem of $\mathfrak{M}$-adequacy where $\mathfrak{M}$ fulfills some special conditions had been the main question in the methodology of sentential logics and it was not generalized before 1958 (J. Łoś and R. Suszko). This generalization will be considered later. Now we will pay attention to the fundamental sets of rules determined by $\langle R, X \rangle$. Namely, we will consider connections between sets of rules determined by propositional system $\langle R, X \rangle$ and by the matrix $\mathfrak{M}$ adequate for $\langle R, X \rangle$. Let us observe, for instance, that

Lemma 2.46. *If $Cn(R, X) = E(\mathfrak{M})$, then*

(i) $\mathrm{Adm}(R, X) = V(\mathfrak{M})$;

(ii) $\mathrm{Der}(R, X) \subseteq V(\mathfrak{M})$;

(iii) $N(\mathfrak{M}) \subseteq \mathrm{Adm}(R, X)$.

These apparent connections need no proofs, and the only thing worth noticing is that the inclusions (ii) and (iii) are not reversible. Completing this observation, let us notice that $\mathfrak{M}$-adequacy of $\langle R, X \rangle$ yields none of the inclusions

$$\mathrm{Der}(R, X) \subseteq N(\mathfrak{M}) \quad , \quad N(\mathfrak{M}) \subseteq \mathrm{Der}(R, X),$$
$$\mathrm{Adm}(R, X) \subseteq N(\mathfrak{M}) \quad , \quad V(\mathfrak{M}) \subseteq \mathrm{Der}(R, X).$$

To show that the first inclusion is not valid for some $\langle R, X \rangle$ and $\mathfrak{M}$, let us take $\langle R_{0*}, A_2 \rangle$ and the two–valued matrix $\mathfrak{M}_2 : r_* \in \mathrm{Der}(R_{0*}, A_2) \setminus N(\mathfrak{M}_2)$. The second inclusion can be refuted with the help of $\mathfrak{M}_{S5}$ and $\langle R_{0a}, Sb(A_{S5}) \rangle$; we have $r_\square \in N(\mathfrak{M}_{S5}) \setminus \mathrm{Der}(R_{0a}, Sb(A_{S5}))$.

Strong adequacy

The above examination shows that it is possible to find some stronger connections between matrices and logics than adequacy. Let us consider the notion of strong adequacy as introduced by [64], 1958:

Definition 2.47. A matrix $\mathfrak{M}$ is *strongly adequate* for a propositional logic $\langle R, X \rangle$ (or the consequence operation Cn_{RX}) iff

$$Cn(R, X \cup Y) = \overrightarrow{\mathfrak{M}}(Y), \quad \text{for all } Y \subseteq S.$$

It is easy to verify that the notion of strong adequacy can also be expressed by the following equality between two sets of rules:

Lemma 2.48 (∞). $\mathfrak{M}$ *is strongly adequate for* $\langle R, X \rangle$ *iff* $N(\mathfrak{M}) = \mathrm{Der}(R, X)$.

To get the proof it suffices to analyze the definitions of the involved notions (see also Lemma 2.33). For finitary, non-axiomatic rules it can be proved that

Lemma 2.48 (fin). $N(\mathfrak{M}) = \mathrm{Der}(R, X)$ *iff* $Cn(R, X \cup Y) = \overrightarrow{\mathfrak{M}}(Y)$ *for every finite non-empty* $Y \subseteq S$.

It is visible that the equivalence 2.48 (∞) is rather some kind of translation of the notion of strong adequacy from the language of formulas into the language of rules. It does not resolve the question whether a given propositional logic has or does not have a strongly adequate matrix.

Definition 2.47 states that $\mathfrak{M}$ is strongly adequate for $\langle R, X \rangle$ iff the consequences $\overrightarrow{\mathfrak{M}}$ and Cn_{RX} are equal. This is the reason (see Corollary 2.24 (iv)) why the investigations concerning the notion are restricted to invariant calculi ($R \subseteq$ Struct and $X = Sb(X)$) or, in other words, to structural consequence operations. Let us note, moreover, that it is possible to give some examples of structural consequences (invariant systems $\langle R, X \rangle$) which do not have any strongly adequate matrix. For instance, consider the pair $\langle \{r\}, \emptyset \rangle$ where r is the rule over S_2 defined by

$$r : \frac{\alpha \to \alpha}{\beta \to \beta} \quad \text{for all } \alpha, \beta \in S_2.$$

Of course, $\langle \{r\}, \emptyset \rangle \in$ Inv. Suppose, on the contrary, that there exists a matrix $\mathfrak{M} = \langle \mathscr{A}, A^* \rangle$ such that $\overrightarrow{\mathfrak{M}}(Y) = Cn(\{r\}, Y)$ for every $Y \subseteq S_2$. From this assumption it follows that

$$q \notin \overrightarrow{\mathfrak{M}}(p \to p) \quad \text{and} \quad q \to q \notin \overrightarrow{\mathfrak{M}}(\emptyset)$$

for some $p, q \in At$ ($p \neq q$). Hence, by the definition of $\overrightarrow{\mathfrak{M}}$, $h_1(p \to p) \in A^*$ and $h_2(q \to q) \notin A^*$ for some homomorphisms h_1 and h_2 from $\mathscr{S}_2$ into $\mathscr{A}$. Take a valuation $v \colon At \to A$ such that $v(p) = h_1(p)$ and $v(q) = h_2(q)$. Then $h^v(p \to p) = h_1(p \to p) \in A^*$ and $h^v(q \to q) = h_2(q \to q) \notin A^*$. Hence $q \to q \notin \overrightarrow{\mathfrak{M}}(p \to p)$, which contradicts our assumptions.

The operation $Cn(X) = \overrightarrow{\mathfrak{M}_3}(X) \cap \overrightarrow{\mathfrak{M}_4}(X)$, where $\mathfrak{M}_i$'s are Łukasiewicz's matrices, is another example of a structural consequence without any strongly adequate matrix. To prove this fact, let us consider two formulas α, β such that $At(\alpha) \cap At(\beta) = \emptyset$ and $\{\alpha\} \in \text{Sat}(\mathfrak{M}_3) \setminus \text{Sat}(\mathfrak{M}_4)$ and $\{\beta\} \in \text{Sat}(\mathfrak{M}_4) \setminus \text{Sat}(\mathfrak{M}_3)$ (e.g., $p \equiv \sim p$ and $q \equiv (q \to \sim q)$). Thus $Cn(\alpha, \beta) = \overrightarrow{\mathfrak{M}_3}(\alpha, \beta) \cap \overrightarrow{\mathfrak{M}_4}(\alpha, \beta) = S_2$, but $Cn(\alpha) = \overrightarrow{\mathfrak{M}_3}(\alpha) \neq S_2$ and $Cn(\beta) = \overrightarrow{\mathfrak{M}_4}(\beta) \neq S_2$. Suppose that $Cn = \overrightarrow{\mathfrak{M}}$ for some matrix $\mathfrak{M}$. Then $\{\alpha, \beta\} \notin \text{Sat}(\mathfrak{M})$ and $\{\alpha\} \in \text{Sat}(\mathfrak{M})$, $\{\beta\} \in \text{Sat}(\mathfrak{M})$, which is impossible since it has been assumed that $At(\alpha) \cap At(\beta) = \emptyset$.

By a similar argument we can show that $\overrightarrow{\mathfrak{M}_n} \cap \overrightarrow{\mathfrak{M}_k}$ is a matrix consequence iff $\overrightarrow{\mathfrak{M}_n} \leqslant \overrightarrow{\mathfrak{M}_k}$ or $\overrightarrow{\mathfrak{M}_k} \leqslant \overrightarrow{\mathfrak{M}_n}$, see J.Hawranek and J.Zygmunt [39], 1981.

Let us also mention that the minimal logic of Johanson as well as Lewis systems $S1-S3$ are also systems without strongly adequate matrices, see J. Hawranek and J. Zygmunt [39], 1981; many facts concerning the notion of strong adequacy can be found in R.Wójcicki [147], 1988.

On the other hand, there are many logics with strongly adequate matrices. First of all,

$$Cn\big(R_0, Sb(A_2) \cup X\big) = \overrightarrow{\mathfrak{M}_2}(X), \quad \text{for every } X \subseteq S_2.$$

Thus, the two-valued matrix $\mathfrak{M}_2$ is strongly adequate for the classical logic. To generalize this fact let us note that for each natural number $n \geqslant 2$,

$$Cn\big(R_0, Sb(\text{Ł}_n) \cup X\big) = \overrightarrow{\mathfrak{M}_n}(X) \quad \text{for each } X \subseteq S_2.$$

This is a simple conclusion from the adequacy theorem for the n-valued Łukasiewicz logic and the deduction theorem for $\overrightarrow{\mathfrak{M}_n}$, see Theorem 1.62.

There is a strongly adequate matrix for the modal system $\langle R_{0a}, Sb(A_{S5})\rangle$ (however, this is not the matrix $\mathfrak{M}_{S5}$) and for every invariant strengthening of the intuitionistic logic $\langle R_0, Sb(A_i)\rangle$. Some general result concerning strong adequacy has been proved in [2], 1982: let X_0 be the set of the following formulas:

$$p \to p$$
$$(p \to q) \to ((q \to s) \to (p \to s))$$
$$(q \to s) \to ((p \to q) \to (p \to s))$$
$$((p \to p) \to (p \to p)) \to (p \to p)$$
$$p \cdot q \to p$$
$$p \cdot q \to q$$
$$(p \to q) \to ((p \to s) \to (p \to q \cdot s))$$
$$p \to p + q$$
$$q \to p + q$$
$$(p \to s) \to ((q \to s) \to (p + q \to s)).$$

Theorem 2.49. *If $\langle R, X \rangle$ is an invariant system over $\mathscr{S}_1$ with $r_0 \in \mathrm{Der}(R, X)$ and $X_0 \subseteq Cn(R, X)$, then there is a matrix $\mathfrak{M}$ strongly adequate for $\langle R, X \rangle$.*

Proof. Let $f: At \to S_1$ be the substitution such that $f(\gamma) = p_0 \to p_0$ for each $\gamma \in At$. From the assumptions it easily follows that

$$(p_0 \to p_0) \to h^f(\alpha), \quad h^f(\alpha) \to (p_0 \to p_0) \in Cn(R, X)$$

for each $\alpha \in S_1$ and hence each formula is satisfiable in the Lindenbaum matrix $\langle \mathscr{S}_1, Cn(R, X \cup Y) \rangle$, where $Y \subseteq S_1$, by the valuation f. Let $\mathfrak{M}$ be the product of the Lindenbaum matrices, i.e.,

$$\mathfrak{M} = \mathop{\mathbf{P}}_{Y \subseteq S} \mathfrak{M}^{R, X \cup Y}.$$

We are going to show that $\overrightarrow{\mathfrak{M}}(Y) = Cn(R, X \cup Y)$ for each $Y \subseteq S_1$. If $\alpha \in \overrightarrow{\mathfrak{M}}(Y)$, then by Theorem 2.41 we obtain $\alpha \in \overrightarrow{\mathfrak{M}^{R, X \cup Y}}(Y)$ hence:

$$h^e(Y) \subseteq Cn(R, X \cup Y) \quad \Rightarrow \quad h^e(\alpha) \in Cn(R, X \cup Y),$$

for each $e: At \to S_1$ which yields $\alpha \in Cn(R, X \cup Y)$.

Assume, on the other hand, that $\alpha \notin \overrightarrow{\mathfrak{M}}(Y)$. Then, also by 2.41, there exists $Z \subseteq S_1$ such that

$$\alpha \notin \overrightarrow{\mathfrak{M}^{R, X \cup Z}}(Y)$$

and therefore

$$h^e(\alpha) \notin Cn(R, X \cup Z) \wedge h^e(Y) \subseteq Cn(R, X \cup Z)$$

for some $e: At \to S_1$. Thus $h^e(\alpha) \notin Cn\big(R, X \cup h^e(Y)\big)$,which implies, on the basis of the invariantness of $\langle R, X \rangle$ (see 1.57), that $\alpha \notin Cn(R, X \cup Y)$. □

A general criterion on the existence of a strongly adequate matrix for a given logic was provided by [64], 1958 (and improved in [140], 1970). This criterion states that the existence of a strongly adequate matrix is equivalent to some kind of separability condition:

Theorem 2.50. *Let Cn be a structural consequence operation (i.e., let $Cn = Cn_{RX}$ for some invariant system $\langle R, X \rangle$). Then there is a matrix strongly adequate for Cn (or $\langle R, X \rangle$) if and only if the following condition is satisfied for all $K \subseteq 2^S$, $Z \subseteq S$ and $\alpha \in S$: if K is a family of Cn-consistent sets such that $At(X) \cap At(Y) = \emptyset$ for all $X, Y \in K$ ($X \neq Y$) and if $At(\bigcup K) \cap At(Z \cup \{\alpha\}) = \emptyset$, then*

$$\alpha \in Cn(\bigcup K \cup Z) \quad \Rightarrow \quad \alpha \in Cn(Z).$$

Corollary 2.51. *Let Cn be a finitistic structural consequence operation. Then there exists a matrix strongly adequate for Cn if and only if*

$$\alpha \in Cn(Z \cup Y) \wedge At(\{\alpha\} \cup Z) \cap At(Y) = \emptyset \quad \Rightarrow \quad \alpha \in Cn(Z)$$

for all $\alpha \in S$, $Z \subseteq S$ and all Cn-consistent $Y \subseteq S$.

As an application of this criterion consider Maksimova's theorem, which states that the separability condition from Corollary 2.51 is satisfied by the relevant logics E and R (see [69], 1976). This leads immediately, by 2.51, to the theorem on the existence of strongly adequate matrices for these logics, see [125], 1979.

Some variants of the matrix semantics

It has been shown that some invariant logics (structural consequences) do not have any strongly adequate matrix. Hence the notion of strong adequacy is not similar to that of $\mathfrak{M}$-adequacy. This fact forced logicians to introduce some variants of the matrix semantics. We will now consider some of them.

Generalized (or ramified matrices) — as introduced in [140], 1970 — are the structures

$$\mathfrak{G}\mathfrak{M} = \langle \mathscr{A}, \{A_t\}_{t\in T}\rangle$$

where $\mathscr{A}$ is an algebra and A_t, for every $t \in T$, is a subset of the universe of $\mathscr{A}$. The consequence operation determined by this structure is

$$\alpha \in \overrightarrow{\mathfrak{G}\mathfrak{M}}(X) \Leftrightarrow \left(h^v(X) \subseteq A_t \Rightarrow h^v(\alpha) \in A_t, \quad \text{for all } t \in T, v\colon At \to A\right).$$

Classes of matrices may also be considered as semantics for propositional logics. For a given class $\mathbf{K}$ of similar matrices one can define the structural consequence $\overrightarrow{\mathbf{K}}$ (see [142], 1973) by

$$\overrightarrow{\mathbf{K}}(X) = \bigcap\{\overrightarrow{\mathfrak{M}}(X) : \mathfrak{M} \in \mathbf{K}\}.$$

In other words, $\overrightarrow{\mathbf{K}} = \prod\{\overrightarrow{\mathfrak{M}} : \mathfrak{M} \in \mathbf{K}\}$. The next theorem is a version of the Lindenbaum result ([142], 1973):

Theorem 2.52. *For every structural consequence operation Cn there is a class $\mathbf{K}$ of matrices (or a generalized matrix $\mathfrak{G}\mathfrak{M}$) such that*

$$Cn = \overrightarrow{\mathbf{K}} \qquad (or \quad Cn = \overrightarrow{\mathfrak{G}\mathfrak{M}}).$$

Proof. Let us consider the class

$$\mathbf{K} = \{\langle \mathscr{S}, Cn(X)\rangle : X \subseteq S\}$$

of Lindenbaum matrices (the so-called *bundle of Lindenbaum matrices*). This class can also be regarded as the generalized matrix

$$\mathfrak{G}\mathfrak{M} = \langle \mathscr{S}, \{Cn(X)\}_{X\subseteq S}\rangle.$$

From structurality of Cn and from the involved definitions it easily follows that $Cn = \overrightarrow{\mathbf{K}} = \overrightarrow{\mathfrak{G}\mathfrak{M}}$. $\square$

Preordered algebras discussed in this chapter can also be regarded as some semantics for propositional logics.

Theorem 2.53 (fin). *If $R \subseteq$ Struct and $\emptyset \neq X = Sb(X)$, then there exists a preordered algebra $\mathfrak{A}$ such that*

$$Cn(R, X \cup Y) = \overrightarrow{\mathfrak{A}}(Y), \quad \text{for each } Y \subseteq S.$$

Proof. Let $\mathfrak{A}_Y = \langle \mathscr{S}, \preccurlyeq_Y \rangle$, for each $Y \subseteq S$, be the preordered algebra determined by the Lindenbaum matrix

$$\mathfrak{M}^{R, X \cup Y} = \langle \mathscr{S}, Cn(R, X \cup Y) \rangle$$

where, for $\alpha, \beta \in S$, we have

$$\alpha \preccurlyeq_Y \beta \quad \Leftrightarrow \quad \alpha \notin Cn(R, X \cup Y) \ \lor \ \beta \in Cn(R, X \cup Y).$$

It has been proved that $\overrightarrow{\mathfrak{A}}_Y = \overrightarrow{\mathfrak{M}^{R, X \cup Y}}$ for each $Y \subseteq S$ and hence, by the structurality of the consequence operation Cn_{RX}, we obtain

$$Cn_{RX} = \prod \{ \overrightarrow{\mathfrak{A}_Y} : Y \subseteq S \}.$$

Let $\mathfrak{A}$ be the product of the family $\{\mathfrak{A}_Y : Y \subseteq S\}$ of preordered algebras. By Theorem 2.29, we have $\overrightarrow{\mathfrak{A}}(Y) \subseteq Cn(R, X \cup Y)$ for each $Y \subseteq S$ and, by Corollary 2.30 (i), we also get $Cn(R, X \cup Y) \subseteq \overrightarrow{\mathfrak{A}}(Y)$ for each $Y \in \text{Fin}(S)$. Since Cn_{RX} is a finitistic consequence operation, then $Cn_{RX}(Y) = \overrightarrow{\mathfrak{A}}(Y)$ for each $Y \subseteq S$. $\square$

Corollary 2.54. *If Cn is a structural, finitistic consequence operation and $Cn(\emptyset)$ is non-empty, then there exists a preordered algebra $\mathfrak{A}$ such that $Cn = \overrightarrow{\mathfrak{A}}$.*

The assumptions that Cn is finitistic and $Cn(\emptyset) \neq \emptyset$ cannot be omitted in the above theorem. We can only replace $Cn(\emptyset) \neq \emptyset$ with $Cn(\emptyset) = \bigcap \{Cn(\alpha) : \alpha \in S\}$.

The consequence $\overrightarrow{\mathfrak{A}}$, where $\mathfrak{A}$ is a preordered algebra, can be viewed as a special case of the consequence determined by a closure operation. Namely let $\mathscr{A}$ be an algebra and let F be an operation on A such that:

(i) $X \subseteq F(X)$,

(ii) $X \subseteq Y \Rightarrow F(X) \subseteq F(Y)$,

(iii) $F\big(F(X)\big) \subseteq F(X)$;   for each $X, Y \subseteq A$.

The pair $\langle \mathscr{A}, F \rangle$ determines a structural consequence operation (see [115], 1962):

$$\alpha \in Cn(X) \quad \Leftrightarrow \quad \big(h^v(\alpha) \in F\big(h^v(X)\big), \quad \text{for each } v \colon At \to A).$$

It can be easily shown that for any structural consequence operation there exists a strongly adequate system $\langle \mathscr{A}, F \rangle$.

Sb-adequacy

Let us observe that the notion of strong adequacy (with respect to logical matrices or generalized matrices or (pre)ordered algebras) has one essential defect. Namely, it is not as general — in the family of propositional logics — as the notion of weak adequacy. This is so because of non-normality of the substitution rule in any non-trivial matrix. This yields immediately that no matrix (generalized matrix ... etc.) is strongly adequate to such a propositional system $\langle R, X \rangle$ that $r_* \in \mathrm{Der}(R, X)$ and $\emptyset \neq Cn(R, X) \neq S$. To eliminate this gap, at least for some logics, we introduce a new notion of Sb-adequacy, similar to the two hitherto assumed and used (i.e., weak adequacy and strong adequacy).

Definition 2.55. A matrix $\mathfrak{M}$ is said to be Sb-adequate with respect to a propositional logic $\langle R, X \rangle$ iff

$$Cn(R, X \cup Y) = \overrightarrow{\mathfrak{M}}(Sb(Y)), \qquad \text{for each } Y \subseteq S.$$

It will be assumed that a rule r is Sb-normal with respect to $\mathfrak{M} = \langle \mathscr{A}, A^* \rangle$, in symbols $r \in QN(\mathfrak{M})$, if and only if

$$h^v(Sb(\Pi)) \subseteq A^* \Rightarrow h^v(\alpha) \in A^*$$

for each sequent $\langle \Pi, \alpha \rangle \in r$ and each valuation $v \colon At \to A$. Of course $r_* \in QN(\mathfrak{M})$ in each matrix $\mathfrak{M}$.

Lemma 2.56. For every matrix $\mathfrak{M}$,

$$QN(\mathfrak{M}) = \bigcap \{ V(\mathfrak{N}) : \mathfrak{N} \subseteq \mathfrak{M} \}.$$

Proof. If $r \in QN(\mathfrak{M})$ and if $\langle \Pi, \alpha \rangle \in r$, then it is easily seen that $\alpha \in \overrightarrow{\mathfrak{M}}(Sb(\Pi))$ and hence, by Theorem 2.38,

$$\alpha \in \bigcap \{ E(\mathfrak{N}) : \mathfrak{N} \subseteq \mathfrak{M} \wedge \Pi \subseteq E(\mathfrak{N}) \}.$$

This proves the inclusion ($\subseteq$). To show the reverse inclusion let us assume that $r \notin QN(\mathfrak{M})$. Then $\alpha \notin \overrightarrow{\mathfrak{M}}(Sb(\Pi))$ for some $\langle \Pi, \alpha \rangle \in r$ and hence, by Theorem 2.38, there exists a submatrix $\mathfrak{N} \subseteq \mathfrak{M}$ such that

$$\alpha \notin E(\mathfrak{N}) \qquad \text{and} \qquad \Pi \subseteq E(\mathfrak{N}).$$

Therefore, $r \notin V(\mathfrak{N})$ for some $\mathfrak{N} \subseteq \mathfrak{M}$. $\qquad \square$

Now we are able to prove the counterpart of Lemma 2.48.

Lemma 2.57 (∞). $\mathfrak{M}$ is Sb-adequate for $\langle R, X \rangle$ $\quad$ iff $\quad QN(\mathfrak{M}) = \mathrm{Der}(R, X)$.

Note that $\mathfrak{M}_2$ is Sb-adequate for $\langle R_{0*}, A_2 \rangle$ and $\mathfrak{M}_2$ is not Sb-adequate with respect to $\langle R_0, Sb(A_2) \rangle$. It also appears that Wajsberg's matrix $\mathfrak{M}_{S5}$ is Sb-adequate for $\langle R_{0a*}, A_{S5} \rangle$, however it is not Sb-adequate for $\langle R_{0*}, A_{S5} \rangle$.

If we consider finite, non-axiomatic rules, then the following counterpart of 2.48 can be proved:

Lemma 2.57 (fin). $QN(\mathfrak{M}) = \mathrm{Der}(R, X)$ *iff* $\overrightarrow{\mathfrak{M}}(Sb(Y)) = Cn(R, X \cup Y)$, *for every non-empty* $Y \in \mathrm{Fin}(S)$.

We also write down a simple corollary of Corollary 2.34:

Corollary 2.58. *Let* $\mathfrak{M}$ *be a finite matrix and let* $E(\mathfrak{M}) \neq \emptyset \neq X$. *Then* $\mathfrak{M}$ *is Sb-adequate for* $\langle R, X \rangle$ *iff* $QN(\mathfrak{M}) = \mathrm{Der}(R, X)$.

The main theorem concerning Sb-adequacy is as follows:

Theorem 2.59. *If* $R \subseteq \mathrm{Struct}$ *and* $X \subseteq S$, *then there is a matrix* $\mathfrak{M}$ *which is Sb-adequate for* $\langle R \cup \{r_*\}, X \rangle$, *i.e.,* $Cn(R \cup \{r_*\}, X \cup Y) = \overrightarrow{\mathfrak{M}}(Sb(Y))$ *for each* $Y \subseteq S$.

Proof. In this proof we will use the following notation: the symbol $S(\alpha)$, for $\alpha \in S$, will denote the sublanguage of S generated by the variables occurring in α, i.e.,

$$\beta \in S(\alpha) \quad \Leftrightarrow \quad At(\beta) \subseteq At(\alpha).$$

The following statements are trivial:

(i) $\beta \in S(\alpha) \quad \Rightarrow \quad S(\beta) \subseteq S(\alpha)$,

(ii) $h^e(S(\alpha)) \subseteq S(h^e(\alpha))$, for every $e \colon At \to S$.

Let us define an operation C over S,

$$\alpha \in C(Y) \quad \Leftrightarrow \quad \alpha \in Cn(R, Sb(X) \cup (Y \cap S(\alpha))),$$

for each $Y \subseteq S$, $\alpha \in S$. We are going to show that C is a structural consequence operation. Let us prove

(iii) $C(Y) \cap S(\alpha) \subseteq Cn(R, Sb(X) \cup (Y \cap S(\alpha)))$, for each $Y \subseteq S$, $\alpha \in S$.

Suppose that $\beta \in C(Y) \cap S(\alpha)$. Then $\beta \in Cn(R, Sb(X) \cup (Y \cap S(\beta)))$ and it means by (i) that $\beta \in Cn(R, Sb(X) \cup (Y \cap S(\alpha)))$.

(iv) $Y \subseteq C(Y)$, for each $Y \subseteq S$.

Let $\alpha \in Y$. Then $\alpha \in Y \cap S(\alpha)$ and hence $\alpha \in Cn(R, Sb(X) \cup (Y \cap S(\alpha)))$.

(v) $Y_1 \subseteq Y_2 \quad \Rightarrow \quad C(Y_1) \subseteq C(Y_2)$, for each $Y_1, Y_2 \subseteq S$.

Assume that $Y_1 \subseteq Y_2$ and let $\alpha \in C(Y_1)$. Then $\alpha \in Cn(R, Sb(X) \cup (Y_1 \cap S(\alpha))) \subseteq$ $\subseteq Cn(R, Sb(X) \cup (Y_2 \cap S(\alpha)))$ and hence $\alpha \in C(Y_2)$.

(vi) $C(C(Y)) \subseteq C(Y)$, for each $Y \subseteq S$.

Suppose that $\alpha \in C(C(Y))$. Then $\alpha \in Cn(R, Sb(X) \cup (C(Y) \cap S(\alpha))) \overset{(iii)}{\subseteq} Cn(R, Sb(X) \cup Cn(R, Sb(X) \cup (Y \cap S(\alpha)))) \subseteq Cn(R, Sb(X) \cup (Y \cap S(\alpha)))$.

(vii) $h^e(C(Y)) \subseteq C(h^e(Y))$, for each $Y \subseteq S$, $e \colon At \to S$.

If $\alpha \in C(Y)$, then $\alpha \in Cn(R, Sb(X) \cup (Y \cap S(\alpha)))$ and from the structurality of the consequence generated by $\langle R, Sb(X)\rangle$ (see Corollary 1.57) it follows that $h^e(\alpha) \in \in Cn(R, Sb(X) \cup h^e(Y \cap S(\alpha))) \overset{(iii)}{\subseteq} Cn(R, Sb(X) \cup (h^e(Y) \cap S(h^e(\alpha))))$.

Statements (iv)–(vii) prove that C is a structural consequence operation. We shall now verify that

(viii) $C(Sb(Y)) = Cn(R, Sb(X) \cup Sb(Y))$, for each $Y \subseteq S$.

The inclusion ($\subseteq$) is obvious. To prove the reverse inclusion let us assume that $\alpha \in Cn(R, Sb(X) \cup Sb(Y))$ and let $e \colon At \to S(\alpha)$ be a substitution such that $e(p) = p$ for all $p \in At(\alpha)$. Then $h^e(\alpha) = \alpha$ and hence by 1.57, we get

$$\alpha \in Cn(R, Sb(X) \cup h^e(Sb(Y))) \subseteq Cn(R, Sb(X) \cup (Sb(Y) \cap h^e(S))).$$

But $e \colon At \to S(\alpha)$ and $S(\alpha)$ is a subalgebra of S; thus $h^e(S) \subseteq S(\alpha)$ and this yields $\alpha \in Cn(R, Sb(X) \cup (Sb(Y) \cap S(\alpha)))$, which was to be proved.

Summarizing: C is a structural consequence operation such that $C(Sb(Y)) = {} = Cn(R \cup \{r_*\}, X \cup Y)$ for each $Y \subseteq S$.

Let $K \subseteq 2^S$, $Z \subseteq S$, $\alpha \in S$ and assume that

$$At(\bigcup K) \cap At(Z \cup \{\alpha\}) = \emptyset.$$

If $\alpha \in C(\bigcup K \cup Z)$, then we also have $\alpha \in Cn(R, Sb(X) \cup (\bigcup K \cup Z) \cap S(\alpha)) = Cn(R, Sb(X) \cup (Z \cap S(\alpha)))$ and hence $\alpha \in C(Z)$.

What we have proved above shows that the separation condition from Theorem 2.50 is satisfied by C. Thus, by 2.50, there exists a matrix $\mathfrak{M}$ such that $\overrightarrow{\mathfrak{M}} = C$ which implies

$$\overrightarrow{\mathfrak{M}}(Sb(Y)) = C(Sb(Y)) = Cn(R \cup \{r_*\}, X \cup Y) \text{ for each } Y \subseteq S,$$

completing the whole proof.   □

Corollary 2.60. *For every structural consequence operation Cn there is a logical matrix $\mathfrak{M}$ such that*

$$Cn \circ Sb = \overrightarrow{\mathfrak{M}} \circ Sb.$$

2.5 Propositional logic and lattice theory

In this section we are going to present the most important facts concerning the relationships between some systems of propositional logics and the theory of implicative lattice with a special stress laid upon the connections between classical logic and Boolean algebras.

Consequences generated by Heyting algebras

Suppose that $\langle A, \cup, \cap \rangle$ is a lattice with $\leqslant$ as the lattice ordering, i.e.,

$$x \leqslant y \quad \Leftrightarrow \quad x \cap y = x, \quad \text{for } x, y \in A.$$

The filter operation $F \colon 2^A \to 2^A$ is defined by

$$F(X) = \{a \in A : a_1 \cap \ldots \cap a_n \leqslant a, \text{ for some } a_1, \ldots, a_n \in X\}.$$

Obviously, for any non-empty $X \subseteq A$, the set $F(X)$ is a filter in $\langle A, \leqslant \rangle$. If A contains the unit element, then $F(\emptyset) = \{1\}$ (see Lemma 2.13 (i)) and $F(\{0\}) = A$ if 0 is the least element in A.

Theorem 1.18 on ultrafilters can be formulated as follows: if $X \subseteq A$, $F(X) \neq A$ and if A contains the least element, then there exists $Y \subseteq A$ such that $X \subseteq Y = F(Y) \neq A$ and $F(Y \cup \{x\}) = A$ for every $x \notin Y$. By Lemma 1.15, we get

Corollary 2.61. *In every Heyting algebra* $\langle A, \dot\to, \cup, \cap, - \rangle$*:*

(i) $a, a \dot\to b \in F(X) \quad \Rightarrow \quad b \in F(X)$;

(ii) $b \in F(X \cup \{a\}) \quad \Leftrightarrow \quad a \dot\to b \in F(X)$;

(iii) $-a \in F(X) \quad \Leftrightarrow \quad F(X \cup \{a\}) = A$.

Let $\mathscr{A}$ be a Heyting (Boolean) algebra. Then the consequence operation determined by $\mathscr{A}$ is defined as follows — see Definition 2.22 —

$$\alpha \in \overrightarrow{\mathscr{A}}(X) \quad \Leftrightarrow \quad \big(h^v(\alpha) \in F\big(h^v(X)\big)\big), \quad \text{for every } v \colon At \to A\big)$$

where $\alpha \in S$, $X \subseteq S$ and $F(\ )$ is the filter operation in $\mathscr{A}$.

Corollary 2.62. *A formula* α *belongs to the set* $\overrightarrow{\mathscr{A}}(X)$ *iff*

$$h^v(X) \subseteq H \quad \Rightarrow \quad h^v(\alpha) \in H$$

for every $v \colon At \to A$ *and every filter* H *in* $\mathscr{A}$.

The set $E(\mathscr{A})$ of $\mathscr{A}$-valid formulas is defined by

$$\alpha \in E(\mathscr{A}) \quad \Leftrightarrow \quad \big(h^v(\alpha) = 1_A, \text{ for every } v \colon At \to A\big).$$

Of course, $E(\mathscr{A}) = \overrightarrow{\mathscr{A}}(\emptyset)$. Note that $\overrightarrow{\mathscr{A}}$ is a structural consequence operation over S_2 (see Corollary 2.24). It has been also shown (see Lemma 2.26) that

Corollary 2.63. *For every pair of Heyting algebras $\mathscr{A}$ and $\mathscr{B}$:*

(i)   *if $\mathscr{A} \cong \mathscr{B}$, then $\overrightarrow{\mathscr{A}} = \overrightarrow{\mathscr{B}}$;*

(ii) *if $\mathscr{B} \subseteq \mathscr{A}$ (or if $\mathscr{B}$ is embeddable in $\mathscr{A}$), then $\overrightarrow{\mathscr{A}} \leqslant \overrightarrow{\mathscr{B}}$.*

The next lemma follows directly from Corollary 2.61.

Lemma 2.64. *For every $X \subseteq S$ and every $\alpha, \beta \in S$:*

(i) $\beta \in \overrightarrow{\mathscr{A}}(X) \quad \Rightarrow \quad (\alpha \to \beta) \in \overrightarrow{\mathscr{A}}(X)$;

(ii) $\beta \in \overrightarrow{\mathscr{A}}(X \cup \{\alpha\}) \quad \Leftrightarrow \quad (\alpha \to \beta) \in \overrightarrow{\mathscr{A}}(X)$.

Therefore, one can easily deduce that $\overrightarrow{\mathscr{A}}(X) = Cn\big(R_0, E(\mathscr{A}) \cup X\big)$ for every finite set $X \subseteq S$. It should be noticed, however, that $\overrightarrow{\mathscr{A}}$ need not be a finitistic operation.

Example. Let us consider the infinite linearly ordered algebra

$$\mathscr{G}_\infty = \langle G_\infty, \dot{\to}, \cup, \cap, - \rangle$$

where $G_\infty = \left\{ \dfrac{1}{n} : n \text{ is a natural number} \right\} \cup \{0\}$ and the operations $\dot{\to}, \cup, \cap, -$ are induced by the natural ordering $\leqslant$ of real numbers, i.e.,

$$x \cup y = \max\{x, y\}$$
$$x \cap y = \min\{x, y\}$$
$$x \dot{\to} y = \begin{cases} 1 & \text{if } x \leqslant y \\ y & \text{if } y < x \end{cases}$$
$$-x = \begin{cases} 1 & \text{if } x = 0 \\ 0 & \text{if } x \neq 0. \end{cases}$$

Suppose that $f \colon G_\infty \to At$ is a one-to-one mapping. We will write p_x (for $x \in G_\infty$) instead of $f(x)$. Let $q \in At$ be a variable such that $q \notin f(G_\infty)$. Take

$$X = \{(p_y \to p_x) \to p_{1/2} : x < y\},$$
$$\alpha = (p_0 \to q) + p_{1/2}.$$

We have to consider, for any valuation $v \colon At \to G_\infty$, the following two cases:

(1) $h^v(p_{1/2}) \in F\big(h^v(X)\big)$,

(2) $h^v(p_{1/2}) \notin F\big(h^v(X)\big)$.

In the first case we get $h^v(\alpha) \in F\big(h^v(X)\big)$. So, we assume (2) and prove

(3) $x < y \Rightarrow v(p_x) < v(p_y)$.

Suppose on the contrary that $v(p_x) \geqslant v(p_y)$ for some $x < y$. Then $h^v(p_y \to p_x) =$
$= v(p_y) \dot{\to} v(p_x) = 1 \in F\big(h^v(X)\big)$. Moreover, $(p_y \to p_x) \to p_{1/2} \in X$ and hence

$$h^v(p_y \to p_x) \dot{\to} h^v(p_{1/2}) \in F\big(h^v(X)\big).$$

Thus, by Corollary 2.61 (i), $h^v(p_{1/2}) \in F\big(h^v(X)\big)$, which contradicts (2).

From (3) it follows that $h^v(\{p_x : x \in G_\infty\})$ is an infinite subset of G_∞ and $h^v(p_0) < h^v(p_x)$ for each $x \in G_\infty$. Consequently, we get $h^v(p_0) = 0$ and hence $h^v(\alpha) = 1 \in F\big(h^v(X)\big)$.

We conclude that $\alpha \in \overrightarrow{\mathscr{G}_\infty}(X)$. Let us prove that $\alpha \notin \overrightarrow{\mathscr{G}_\infty}(Y)$ for each finite set $Y \subseteq X$. Assume $Y \in \mathrm{Fin}(X)$. Then $At(Y)$ is also finite and hence there is a natural number n such that $At(Y) \subseteq \{p_0\} \cup \left\{p_x : x \geqslant \dfrac{1}{n}\right\}$. Take $v \colon At \to G_\infty$ such that $v(p_x) = x$ if $x \neq 0$, $v(p_0) = \dfrac{1}{n+1}$ and $v(q) = 0$. Then

$$h^v\big((p_y \to p_x) \to p_{1/2}\big) = 1 \quad \text{if } 0 < x < y.$$

Thus, to get $h^v(Y) \subseteq \{1\}$ it suffices to show $h^v\big((p_y \to p_0) \to p_{1/2}\big) = 1$ if $y \geqslant \dfrac{1}{n}$.

Let us suppose that $y \geqslant \dfrac{1}{n}$. Then $h^v(p_y \to p_0) = y \dot{\to} \dfrac{1}{n+1} = \dfrac{1}{n+1}$ and hence $h^v\big((p_y \to p_0) \to p_{1/2}\big) = \dfrac{1}{n+1} \dot{\to} \dfrac{1}{2} = 1$. Therefore, we have $h^v(\alpha) = \dfrac{1}{2} \notin F(\{1\}) = F\big(h^v(Y)\big)$.

We can only state, on the basis of Theorem 2.27, that

Corollary 2.65. *If $\mathscr{A}$ is finite, then $\overrightarrow{\mathscr{A}}$ is a finitistic consequence operation.*

Therefore, $\overrightarrow{\mathscr{A}}(X) = Cn\big(R_0, E(\mathscr{A}) \cup X\big)$ for each $X \subseteq S$ if $\mathscr{A}$ is finite.

The algebra $\mathscr{A}$ determines, as a (pre)ordered algebra, the two sets of rules; rules normal in $\mathscr{A}$, denoted by $N(\mathscr{A})$, and rules valid in $\mathscr{A}$, denoted by $V(\mathscr{A})$:

$$r \in N(\mathscr{A}) \quad \Leftrightarrow \quad \big(h^v(\alpha) \in F(h^v(\Pi)), \ \text{for every } \langle \Pi, \alpha \rangle \in r, \ v \colon At \to A\big),$$
$$r \in V(\mathscr{A}) \quad \Leftrightarrow \quad \big(\Pi \subseteq E(\mathscr{A}) \Rightarrow \alpha \in E(\mathscr{A}), \ \text{for every } \langle \Pi, \alpha \rangle \in r\big).$$

General statements concerning the two above notions can be found in Section 2.2. Let us note additionally that

Lemma 2.66. *The following conditions are equivalent:*

(i) *r is normal in $\mathscr{A}$ (i.e., $r \in N(\mathscr{A})$);*

(ii) *$h^v(\Pi) \subseteq H \Rightarrow h^v(\alpha) \in H$, for every $\langle \Pi, \alpha \rangle \in r$, $v \colon At \to A$ and filter H;*

(iii) *$\alpha \in \overrightarrow{\mathscr{A}}(\Pi)$, for every $\langle \Pi, \alpha \rangle \in r$.*

From Corollary 2.61 (i) it follows that the modus ponens rule is normal in $\mathscr{A}$. This can be strengthened by proving that

(fin) $$N(\mathscr{A}) = \mathrm{Der}\big(R_0, E(\mathscr{A})\big).$$

The above equality concerns only rules with finite sets of premises. Evidently, the same will be true for arbitrary rules if $\overrightarrow{\mathscr{A}}$ is finitistic; for instance, if $\mathscr{A}$ is finite.

Lemma 2.67. *If H is a filter in a Heyting algebra $\mathscr{A}$, then $\overrightarrow{\mathscr{A}} \leqslant \overrightarrow{\mathscr{A}/H}$.*

Proof. Suppose that $\alpha \notin \overrightarrow{\mathscr{A}/H}(X)$. Then there is a valuation $w\colon At \to A/H$ and a filter H_1 in $\mathscr{A}/H$ (see Corollary 2.62) such that $h^w(X) \subseteq H_1$ and $h^w(\alpha) \notin H_1$.

Let $h\colon A \to A/H$ be the canonical homomorphism. Using the Axiom of Choice we can choose a valuation $v\colon At \to A$ such that $h \circ v = w$. Thus, we get $h \circ h^v = h^w$, which gives $h^v(X) \subseteq h^{-1}(H_1)$ and $h^v(\alpha) \notin h^{-1}(H_1)$. But the set $h^{-1}(H_1)$ is a filter in $\mathscr{A}$ — see Theorem 1.16 — and consequently $\alpha \notin \overrightarrow{\mathscr{A}}(X)$ by Corollary 2.62. $\qquad\square$

Since each proper filter is contained in a prime one, see Theorem 1.20, we get

Corollary 2.68. *If $\mathscr{A}$ is a Heyting algebra and $X \subseteq S$, then*

$$\overrightarrow{\mathscr{A}}(X) = \bigcap \{\overrightarrow{\mathscr{A}/H}(X) : H \ \ \text{is a prime filter in } \ \mathscr{A}.\}$$

Proof. The inclusion ($\subseteq$) follows from Lemma 2.67. To prove ($\supseteq$) let $\alpha \notin \overrightarrow{\mathscr{A}}(X)$. Then, by 1.20 and 2.62, there exist a valuation $v\colon At \to A$ and a prime filter H such that $h^v(X) \subseteq H$ and $h^v(\alpha) \notin H$. Let $h\colon At \to A/H$ be the canonical homomorphism. Then $h\big(h^v(X)\big) \subseteq \{1_{A/H}\}$ and $h\big(h^v(\alpha)\big) \neq 1_{A/H}$, hence $\alpha \notin \overrightarrow{\mathscr{A}/H}(X)$. $\qquad\square$

So, we get $\overrightarrow{\mathscr{A}} = \prod\{\overrightarrow{\mathscr{A}/H} : H$ is a prime filter in $\mathscr{A}\}$.

Theorem 2.69. *Let $\mathscr{A}$ be a Heyting algebra. Then $\alpha \in \overrightarrow{\mathscr{A}}\big(Sb(X)\big)$ if and only if*

$$X \subseteq E(\mathscr{B}) \quad \Rightarrow \quad \alpha \in E(\mathscr{B})$$

for each filter H in $\mathscr{A}$ and each $\mathscr{B} \subseteq \mathscr{A}/H$.

Proof. We get one implication by 2.67 and 2.63 (ii). Now, assume $\alpha \notin \overrightarrow{\mathscr{A}}\big(Sb(X)\big)$. Then, by 2.62, there exist a valuation $v\colon At \to A$ and a filter H in $\mathscr{A}$ such that $h^v\big(Sb(X)\big) \subseteq H$ and $h^v(\alpha) \notin H$. Let $\mathscr{B}$ be the subalgebra of $\mathscr{A}/H$ generated by the set $h\big(h^v(S_2)\big)$, where $h\colon A \to A/H$ is the canonical homomorphism. Since the composition $h \circ h^v$ is a homomorphism from $\mathscr{S}_2$ into $\mathscr{A}/H$, we conclude that $h\big(h^v(S_2)\big)$ is closed under the operations of $\mathscr{A}/H$ (see Lemma 1.2) and hence $h\big(h^v(S_2)\big)$ is the universe of $\mathscr{B}$, i.e., $B = h\big(h^v(S_2)\big)$.

We shall prove that $X \subseteq E(\mathcal{B})$. Let $w \colon At \to h(h^v(S_2))$ be any valuation. Using the Axiom of Choice we can define a substitution $e \colon At \to S_2$ such that $w = h \circ h^v \circ e$. Thus $h^w(X) = h(h^v(h^e(X))) \subseteq h(h^v(Sb(X))) \subseteq h(H) \subseteq \{1_{A/H}\}$.

We get $h^w(X) \subseteq \{1_{A/H}\}$ for every $w \colon At \to h(h^v(S))$, hence $X \subseteq E(\mathcal{B})$. Moreover, observe that $h \circ h^v \colon S_2 \to h(h^v(S_2))$ and $h \circ h^v(\alpha) \neq 1_{A/H}$. Thus, $X \subseteq E(\mathcal{B})$ and $\alpha \notin E(\mathcal{B})$, which was to be proved. $\qquad\square$

Since the product of Heyting algebras coincides with the product of ordered algebras as introduced in Section 2.2, we will use in our considerations some results obtained there. For instance, it follows from Corollary 2.30 that

Corollary 2.70. *For each family* $\{\mathscr{A}_t\}_{t \in T}$ *of Heyting algebras:*

(i) $E(\underset{t \in T}{\mathbf{P}} \mathscr{A}_t) = \bigcap\{E(\mathscr{A}_t) : t \in T\}$;

(ii) *if* T *is finite, then* $\overrightarrow{\underset{t \in T}{\mathbf{P}} \mathscr{A}}_t(X) = \bigcap\{\overrightarrow{\mathscr{A}_t}(X) : t \in T\}$, *for each* $X \subseteq S$.

The filter consequence operations

In terms of lattice theory we define three consequence operations FC_i, FC_l and FC_2 over the language $\mathscr{S}_2$ and then examine connections between the introduced operations and some propositional systems. In result, we prove completeness theorems for the intuitionistic logic, the linear logic of Dummett and the classical propositional logic.

(i) The set $FC_i(X)$, for $X \subseteq S_2$, is the set of those formulas $\alpha \in S_2$ which in any Heyting algebra $\mathscr{A}$ fulfill the condition

$$h^v(\alpha) \in F(h^v(X)), \quad \text{for every } v \colon At \to A.$$

(ii) The set $FC_l(X)$, for $X \subseteq S_2$, is the set of those formulas $\alpha \in S_2$ which in any linear Heyting algebra $\mathscr{A}$ fulfill the condition

$$h^v(\alpha) \in F(h^v(X)), \quad \text{for every } v \colon At \to A.$$

(iii) The set $FC_2(X)$, for $X \subseteq S_2$, is the set of those formulas $\alpha \in S_2$ which in any Boolean algebra $\mathscr{A}$ fulfill the condition

$$h^v(\alpha) \in F(h^v(X)), \quad \text{for every } v \colon At \to A.$$

Further results containing symbol FC will concern FC_i, FC_l as well as FC_2, simultaneously. Directly from the definitions it follows that

Lemma 2.71. *For every* $X, Y \subseteq S_2$:

(i) $X \subseteq FC(X)$,

(ii) $X \subseteq Y \Rightarrow FC(X) \subseteq FC(Y)$,

(iii) $FC\big(FC(X)\big) \subseteq FC(X)$,

(iv) $h^e\big(FC(X)\big) \subseteq FC\big(h^e(X)\big)$ *for all* $e \colon At \to S_2$.

Lemma 2.72. $FC(X) \subseteq \bigcup\{FC(Y) : Y \in \mathrm{Fin}(X)\}$ *for every* $X \subseteq S_2$.

Proof. Assume that $\alpha \notin FC(Y)$ for each finite $Y \subseteq X$. Therefore, with each $Y \in \mathrm{Fin}(X)$ there can be associated Heyting (or linear or Boolean) algebra $\mathscr{A}_Y$, homomorphism $h_Y \colon \mathscr{S}_2 \to \mathscr{A}_Y$ and filter H_Y in $\mathscr{A}_Y$ such that

$$h_Y(Y) \subseteq H_Y \quad \text{and} \quad h_Y(\alpha) \notin H_Y.$$

Let $\mathscr{A}$ be the product of the algebras $\mathscr{A}_Y$.

Define the homomorphism $h \colon \mathscr{S}_2 \to \mathscr{A}$ as in Lemma 1.6, i.e.,

$$h(\varphi) = \langle h_Y(\varphi)\rangle_{Y \in \mathrm{Fin}(X)} \quad \text{for all} \ \ \varphi \in S_2.$$

We have to prove that $h(\alpha) \notin \mathbf{F}\big(h(X)\big)$ where $\mathbf{F}$ is the filter-operation in $\mathscr{A}$. Suppose that $h(\alpha) \in \mathbf{F}\big(h(X)\big)$. Then, by 2.12 (iv), $h(\alpha) \in \mathbf{F}\big(h(Y)\big)$ for some $Y \in \mathrm{Fin}(X)$ and hence $h_Y(\alpha) = \pi_Y\big(h(\alpha)\big) \in \pi_Y\big(\mathbf{F}\big(h(Y)\big)\big) \subseteq H_Y$ against our assumptions. Thus, $\alpha \notin FC(X)$ when FC is FC_i or FC_2.

Now let us assume that $FC = FC_l$. The algebra $\mathscr{A}$ constructed above is not linear, but by Corollary 2.70 (i), $(p \to q) + (q \to p) \in \bigcap\{E(\mathscr{A}_Y) : Y \in \mathrm{Fin}(X)\} = E(\mathscr{A})$.

Moreover, we know that $\alpha \notin \overrightarrow{\mathscr{A}}(X)$ and hence by Corollary 2.68, there is a prime filter H in $\mathscr{A}$ such that $\alpha \notin \overrightarrow{\mathscr{A}/H}(X)$. It follows now from 1.19 and 2.67 that $\mathscr{A}/H$ is prime and $(p \to q) + (q \to p) \in E(\mathscr{A}/H)$. Hence $x \overrightarrow{\to} y = 1$ or $y \overrightarrow{\to} x = 1$ for each x, y in $\mathscr{A}/H$. Then $\mathscr{A}/H$ is a linear algebra and $\alpha \notin \overrightarrow{\mathscr{A}/H}(X)$. Thus $\alpha \notin FC_l(X)$. $\qquad\square$

By Lemma 2.72, FC is a finitistic consequence operation. Moreover, as a simple conclusion from the definitions of FC and Corollary 2.61 we get

Corollary 2.73. *For each* $X \subseteq S_2$ *and each* $\alpha, \beta \in S_2$:

(i) $\beta \in FC(X) \quad \Rightarrow \quad (\alpha \to \beta) \in FC(X)$;

(ii) $\beta \in FC(X \cup \{\alpha\}) \quad \Leftrightarrow \quad (\alpha \to \beta) \in FC(X)$;

(iii) $\alpha \, , \ \alpha \to \beta \in FC(X) \quad \Rightarrow \quad \beta \in FC(X)$;

(iv) $Sb\big(FC(\emptyset)\big) \subseteq FC(\emptyset)$.

Intuitionistic propositional logic

The system $\langle R_0, Sb(A_i)\rangle$ of intuitionistic logic is introduced in Section 1.5. We also recall that the filter consequence FC_i is defined as

$$\alpha \in FC_i(X) \quad \Leftrightarrow \quad (\alpha \in \overrightarrow{\mathscr{A}}(X), \text{ for each Heyting algebra } \mathscr{A}).$$

From among all formulas of $\mathscr{S}_2$ the subset $FC_i(\emptyset)$ is separated:

Corollary 2.74. $FC_i(\emptyset)$ *is the set of all those formulas* $\alpha \in \mathscr{S}_2$ *which in each Heyting algebra* $\mathscr{A}$ *fulfill the condition*

$$h^v(\alpha) = 1_A, \quad \text{for all} \;\; v \colon At \to A.$$

The symbol A_i has been previously used to denote the axioms of intuitionistic logic. Evidently we obtain

Lemma 2.75. $A_i \subseteq FC_i(\emptyset)$.

The standard proof of the completeness theorem for the intuitionistic logic is well-known. It is based on the construction of the Lindenbaum–Tarski algebra which appears to be a (free) Heyting algebra.

Theorem 2.76. $Cn\big(R_0, Sb(A_i) \cup X\big) = FC_i(X)$, *for each* $X \subseteq S_2$.

The relationships between intuitionistic logic and Heyting algebras were stated by many authors — an exhaustive exposition of this subject is presented in the monograph [107], 1963. It should be noticed that in Theorem 2.76 the stress is laid on the relationships between the syntactic notion of consequence operation based upon the intuitionistic logic and the algebraic notion of filter consequence. Obviously, as an immediate result of 2.76 we get $Cn(R_{0*}, A_i) = FC_i(\emptyset)$, i.e., the set $Cn(R_{0*}, A_i)$ consists of all formulas which are valid in every Heyting algebra (all intuitionistic tautologies). By Theorems 2.76 and 2.69, we get

Corollary 2.77. *For each* $\alpha \in S_2$ *and each* $X \subseteq S_2$,

$$\alpha \in Cn(R_{0*}, A_i \cup X) \Leftrightarrow \big(X \subseteq E(\mathscr{A}) \Rightarrow \alpha \in E(\mathscr{A}), \text{for each Heyting algebra } \mathscr{A}\big)$$

The above results allow us to characterize rules derivable in intuitionistic logic. Namely, a rule r is derivable in $\langle R_{0*}, A_i\rangle$ (i.e., $r \in \mathrm{Der}(R_{0*}, A_i)$) iff r is valid in each Heyting algebra $\mathscr{A}$ (i.e., $r \in V(\mathscr{A})$). For the invariant system $\langle R_0, Sb(A_i)\rangle$ one can prove that the following conditions are equivalent:

(i) $r \in \mathrm{Der}\big(R_0, Sb(A_i)\big)$;

(ii) $r \in N(\mathscr{A})$, for each Heyting algebra $\mathscr{A}$;

(iii) $\big(h^v(X) \subseteq \{1_A\} \Rightarrow h^v(\alpha) = 1_A\big)$, for each $\langle X, \alpha\rangle \in r$ and every valuation v in each Heyting algebra $\mathscr{A}$.

Now, we will try to restrict the class of intuitionistic models (i.e., the class of all Heyting algebras) to the family of the Jaśkowski algebras $\{\mathscr{J}_n\}_{n\geqslant1}$. First, we shall prove (by methods from [107], 1963)

Theorem 2.78. *For each* $\alpha \in S_2$ *the following conditions are equivalent:*

(i) $\alpha \in Cn(R_{0*}, A_i)$;

(ii) α *is valid (i.e.,* $\alpha \in E(\mathscr{A})$*) in all finite Heyting algebras.*

Proof. The implication (i)$\Rightarrow$(ii) is obvious. For the proof of (ii)$\Rightarrow$(i) we assume that $\alpha \notin Cn(R_{0*}, A_i)$. Then, by Theorem 2.76, there are a Heyting algebra $\mathscr{A} = \langle A, \dot\rightarrow, \cup, \cap, - \rangle$ and a valuation $v \colon At \to A$ such that $h^v(\alpha) \neq 1_A$.

Let $\langle K, \cup, \cap \rangle$ be the sublattice of $\langle A, \cup, \cap \rangle$ generated by $\{1_A\} \cup h^v\big(Sf(\alpha)\big)$ (recall that $Sf(\alpha)$ denotes the set of all subformulas of α, and notice that the set $Sf(\alpha)$ is finite). Then, by Lemma 1.14, $\langle K, \cup, \cap \rangle$ is a finite distributive lattice with the lattice ordering $\leqslant$ induced from $\mathscr{A}$, i.e.,

$$x \leqslant y \quad \Leftrightarrow \quad x \leqslant_A y, \quad \text{for all } x, y \in K.$$

But every distributive lattice ordering on a finite set is a Heyting ordering, hence we can define the operations $\dot\rightarrow_1, -_1$ in such a way that $\mathscr{K} = \langle K, \dot\rightarrow_1, \cup, \cap, -_1 \rangle$ will be the Heyting algebra determined by $\leqslant$. Since, on the other hand, the relation $\leqslant$ is induced from $\mathscr{A}$, there are some connections between the operations $\dot\rightarrow, -$ and $\dot\rightarrow_1, -_1$. For instance, it can be easily shown that for $x, y \in K$:

$$
\begin{array}{llll}
\text{if} & x \dot\rightarrow y \ \in K, & \text{then} & x \dot\rightarrow_1 y \ = x \dot\rightarrow y; \\
\text{if} & -x \ \in K, & \text{then} & -_1 x \ = -x.
\end{array}
$$

Let us consider a homomorphism $h \colon \mathscr{S}_2 \to \mathscr{K}$ fulfilling the condition

$$h(\gamma) = v(\gamma), \quad \text{for every } \gamma \in At(\alpha).$$

We easily show, by induction on the length of a formula, that $h(\varphi) = h^v(\varphi)$ for each $\varphi \in Sf(\alpha)$. This follows immediately from the fact that $\langle K, \cup, \cap \rangle$ is a sublattice of $\langle A, \cup, \cap \rangle$ and the above mentioned connections between the operations in $\mathscr{K}$ and $\mathscr{A}$. Therefore

$$h(\alpha) = h^v(\alpha) \neq 1_A = 1_K$$

and then α is not valid in the finite Heyting algebra $\mathscr{K}$. $\square$

The next theorem states that Jaśkowski algebras $\{\mathscr{J}_n\}_{n\geqslant1}$ constitute an adequate family of algebras for the intuitionistic propositional logic:

Theorem 2.79. $Cn(R_{0*}, A_i) = \bigcap\{E(\mathscr{J}_n) : n \geqslant 1\}$.

Proof. We have to prove the inclusion ($\supseteq$) only. Assume that $\alpha \notin Cn(R_{0*}, A_i)$. Then, by 2.78, there is a finite Heyting algebra $\mathscr{A}$ such that $\alpha \notin E(\mathscr{A})$. Since $\mathscr{A}$ is finite, $\mathscr{A}$ is embeddable in algebra $(\mathscr{J}_n)^m$ for some natural numbers n, m

(see Corollary 1.30). It follows then from Corollaries 2.70 (ii) and 2.63 (ii) that $\overrightarrow{\mathscr{J}_n} = \overrightarrow{(\mathscr{J}_n)^m} \leqslant \overrightarrow{\mathscr{A}}$ and hence $E(\mathscr{J}_n) \subseteq E(\mathscr{A})$, which means that $\alpha \notin E(\mathscr{J}_n)$ for some $n \geqslant 1$. □

Using the deduction theorem Corollary 2.73 (ii) we can strengthen the above result to

Corollary 2.80. *For every finite set* $X \subseteq S_2$,

$$Cn\big(R_0, Sb(A_i) \cup X\big) = \bigcap\{\overrightarrow{\mathscr{J}_n}(X) : n \geqslant 1\}.$$

Theorem 2.79 was first stated by S. Jaśkowski [44], 1936. We have presented here another proof of this fact based on certain results (e.g., Theorem 2.78) established by H. Rasiowa and R. Sikorski [107], 1963.

It is worth noticing that the equality stated in Corollary 2.80 does not hold for arbitrary (non-finite) sets X. Using some results from [148], 1973, we can even prove that there is a finite set X such that $Cn(R_{0*}, A_i \cup X) \neq \bigcap\{\overrightarrow{\mathscr{J}_n}(Sb(X)) : n \geqslant 1\}$. Therefore, $\mathrm{Der}(R_{0*}, A_i) \neq \bigcap\{V(\mathscr{J}_n) : n \geqslant 1\}$. Corollary 2.80 allows us only to characterize derivability in $\langle R_0, Sb(A_i)\rangle$ for rules with finite sets of premises. Namely, it is easy to observe that such a rule r belongs to $\mathrm{Der}\big(R_0, Sb(A_i)\big)$ if and only if r is normal in each algebra $\mathscr{J}_n$.

Linear propositional logic of Dummett

The system $\langle R_0, Sb(A_l)\rangle$ of *Dummett's linear logic* is formalized in the language $\mathscr{S}_2$. Let us define

$$A_l = A_i \cup \{(p \rightarrow q) + (q \rightarrow p)\}.$$

The filter consequence connected with this logic is FC_l, where $\alpha \in FC_l(X)$ means that $\alpha \in \overrightarrow{\mathscr{A}}(X)$, for all linear Heyting algebras $\mathscr{A}$.

Corollary 2.81. $FC_l(\emptyset)$ *is the set of those formulas* $\alpha \in S_2$ *which in each linear algebra* $\mathscr{A}$ *fulfill the condition* $h^v(\alpha) = 1_A$ *for all* $v \colon At \rightarrow A$.

Corollary 2.82. $A_l \subseteq FC_l(\emptyset)$.

Note that $FC_i(\emptyset) \neq FC_l(\emptyset)$ since the formula $(p \rightarrow q) + (q \rightarrow p)$ is not valid in $\mathscr{J}_3$. We shall prove the following completeness theorem for Dummett's system:

Theorem 2.83. $Cn\big(R_0, Sb(A_l) \cup X\big) = FC_l(X)$, *for each* $X \subseteq S_2$.

Proof. The inclusion ($\subseteq$) follows from 2.82 and 2.73. To prove the reverse inclusion let us assume that $\alpha \notin Cn\big(R_0, Sb(A_l)\big)$. Then, by Corollary 2.77,

$$(p \rightarrow q) + (q \rightarrow p) \in E(\mathscr{A}) \quad \text{and} \quad \alpha \notin E(\mathscr{A})$$

for some Heyting algebra $\mathscr{A}$. Thus, by Corollary 2.68, there exists a prime filter H in $\mathscr{A}$ such that

$$(p \rightarrow q) + (q \rightarrow p) \in E(\mathscr{A}/H) \quad \text{and} \quad \alpha \notin E(\mathscr{A}/H).$$

Since the algebra $\mathscr{A}/H$ is prime (see Theorem 1.19) and $(a \dot\rightarrow b) \cup (b \dot\rightarrow a) = 1$ for each a and b in A/H, we conclude that $\mathscr{A}/H$ is linearly ordered. Thus, $\alpha \notin FC_l(\emptyset)$.

It has been proved that $FC_l(\emptyset) = Cn(R_0, Sb(A_l))$. To prove the inclusion $FC_l(X) \subseteq Cn(R_0, Sb(A_l) \cup X)$ for each X, it suffices now to use the finiteness of the consequence FC_l and the deduction theorem (see Lemma 2.72 and Corollary 2.73 (ii)).

In fact, if $\alpha \in FC_l(X)$, then $\alpha_1 \rightarrow (\ldots(\alpha_n \rightarrow \alpha)\ldots) \in FC_l(\emptyset) = Cn(R_0, Sb(A_l))$ for some $\alpha_1, \ldots, \alpha_n \in X$. Thus $\alpha \in Cn(R_0, Sb(A_l) \cup X)$. $\quad\square$

Now we will try to restrict the family of all linear algebras to Gödel–Heyting's algebras $\{\mathscr{G}_n\}_{n \geqslant 1}$ defined in Section 1.3.

Lemma 2.84. *If $\mathscr{A}$ is an infinite linear Heyting algebra, then*

$$E(\mathscr{A}) = \bigcap\{E(\mathscr{G}_n) : n \geqslant 1\}.$$

Proof. The inclusion ($\subseteq$) follows from Lemma 1.31 and Corollary 2.63 (ii). To prove the inclusion ($\supseteq$) let us assume that $\alpha \notin E(\mathscr{A})$. Then $h^v(\alpha) \neq 1_A$ for some valuation v in A. Since $\mathscr{A}$ is linearly ordered, each subset of A containing 1_A and 0_A is closed under the operations of $\mathscr{A}$, i.e., is a subalgebra. The set $h^v(At(\alpha)) \cup \{1_A, 0_A\}$ determines thus a finite subalgebra $\mathscr{B}$ of $\mathscr{A}$ such that $\alpha \notin E(\mathscr{B})$. By Corollary 1.32, $\mathscr{B}$ is isomorphic with some algebra $\mathscr{G}_n$ and hence $\alpha \notin E(\mathscr{G}_n)$. $\quad\square$

On the basis of Lemma 2.84 and Theorem 2.83 we conclude that each infinite linear Heyting algebra $\mathscr{A}$ (e.g., algebra $\mathscr{G}_\infty$) is adequate (weakly) for Dummett's system, i.e.,

$$Cn(R_{0*}, A_l) = E(\mathscr{A}) = \bigcap\{E(\mathscr{G}_n) : n \geqslant 1\}.$$

Moreover, observe that 2.84 and the deduction theorem allow us to prove

$$Cn(R_0, Sb(A_l) \cup X) = \overrightarrow{\mathscr{G}_\infty}(X) = \bigcap\{\overrightarrow{\mathscr{G}_n}(X) : n \geqslant 1\} = \overrightarrow{\mathscr{A}}(X)$$

for each finite set X and each infinite algebra $\mathscr{A}$. Hence,

$$r \in \mathrm{Der}(R_0, Sb(A_l)) \quad \Leftrightarrow \quad (r \in N(\mathscr{G}_n) \text{ for each } n \geqslant 2).$$

It should be noticed, however, that this equivalence concerns only rules with finite sets of premises. It is easy to observe that the above equalities do not hold for any (infinite) sets X. In order to show that $Cn(R_0, Sb(A_l) \cup X) \neq \overrightarrow{\mathscr{G}_\infty}(X)$ for some X it suffices to make a use of the example on page 76 where it has been shown that the consequence $\overrightarrow{\mathscr{G}_\infty}$ is not finitistic. Observe, moreover, that $Cn(R_0, Sb(A_l) \cup X) \neq \bigcap\{\overrightarrow{\mathscr{G}_n}(X) : n \geqslant 1\}$ for the same set X. For arbitrary (finite or infinite) rules we can only prove the equivalence of the following conditions:

(i) $r \in \mathrm{Der}(R_0, Sb(A_l))$;

(ii) $r \in N(\mathscr{A})$ in each linear algebra $\mathscr{A}$;

(iii) $h^v(X) \subseteq \{1_A\} \Rightarrow h^v(\alpha) = 1_A$, for each $\langle X, \alpha \rangle \in r$ and each valuation v in each linearly ordered algebra $\mathscr{A}$.

The counterpart of Theorem 2.83 for the system $\langle R_{0*}, A_l \rangle$ will be as follows:

Theorem 2.85. *For each* $X, \alpha \subseteq S_2$ *the following conditions are equivalent:*

(i) $\alpha \in Cn\big(R_{0*}, A_l \cup X\big)$;

(ii) $X \subseteq E(\mathscr{A}) \Rightarrow \alpha \in E(\mathscr{A})$, *for each linear algebra* $\mathscr{A}$;

(iii) $\alpha \in \overrightarrow{\mathscr{A}}\big(Sb(X)\big)$, *for some infinite linear algebra* $\mathscr{A}$;

(iv) $X \subseteq E(\mathscr{G}_n) \Rightarrow \alpha \in E(\mathscr{G}_n)$, *for each* $n \geqslant 1$.

Proof. The implication (i)$\Rightarrow$(iii) is obvious — see Theorem 2.83.

(iii)$\Rightarrow$(iv): By Lemma 1.31, $\mathscr{G}_n$ is embeddable in $\mathscr{A}$ and hence $\overrightarrow{\mathscr{A}} \leqslant \overrightarrow{\mathscr{G}_n}$ (see Corollary 2.63 (ii)) which means in particular that $\overrightarrow{\mathscr{A}}\big(E(\mathscr{G}_n)\big) \subseteq E(\mathscr{G}_n)$. So, if $\alpha \in \overrightarrow{\mathscr{A}}\big(Sb(X)\big)$ and if $X \subseteq E(\mathscr{G}_n)$, then also $\alpha \in E(\mathscr{G}_n)$.

(iv)$\Rightarrow$(ii): According to Corollary 1.32 (iii) each finite linearly ordered algebra $\mathscr{A}$ is isomorphic with some algebra $\mathscr{G}_n$ and hence $E(\mathscr{A}) = E(\mathscr{G}_n)$. Thus, if we assume (iv), then

$$X \subseteq E(\mathscr{A}) \quad \Rightarrow \quad \alpha \in E(\mathscr{A})$$

for each finite linearly ordered algebra $\mathscr{A}$. Moreover, from (iv) it results that

$$\text{if } X \subseteq E(\mathscr{G}_n) \text{ for each } n, \text{ then } \alpha \in E(\mathscr{G}_n) \text{ for each } n.$$

By Lemma 2.84, the condition (ii) is fulfilled for each infinite linear algebra $\mathscr{A}$.

(ii)$\Rightarrow$(i): Assume (ii) for some X, α. Since each subalgebra and each quotient algebra of a linear Heyting algebra $\mathscr{A}$ is also a linear Heyting algebra, we conclude by Theorem 2.69 that $\alpha \in \overrightarrow{\mathscr{A}}\big(Sb(X)\big)$. Thus, by 2.83, $\alpha \in Cn(R_{0*}, A_l \cup X)$. $\square$

From the above theorem it immediately follows that for each set $X \subseteq S_2$:

$$Cn(R_{0*}, A_l \cup X) = \begin{cases} E(\mathscr{G}_\infty) & \text{if } X \subseteq E(\mathscr{G}_\infty) \\ E(\mathscr{G}_n) & \text{if } X \subseteq E(\mathscr{G}_n) \wedge X \setminus E(\mathscr{G}_{n+1}) \neq \emptyset. \end{cases}$$

Moreover, we conclude that $r \in \text{Der}(R_{0*}, A_l)$ provided that r is valid in each algebra $\mathscr{G}_n$, i.e., $r \in V(\mathscr{G}_n)$ for each n. Obviously, each rule derivable in $\langle R_{0*}, A_l \rangle$ is valid in each linear algebra.

Some of the above results can also be used to characterize the so-called Gödel–Heyting's propositional logic, i.e., the logics determined by the algebras $\{\mathscr{G}_n\}_{n \geqslant 1}$. Let us prove that the consequences $\overrightarrow{\mathscr{G}}_n$, $\overrightarrow{\mathscr{G}}_n \circ Sb$ can be axiomatized with the help of the formulas φ_n defined as

$$\varphi_1 = (p_1 \equiv p_2),$$
$$\varphi_{n+1} = \varphi_n + (p_1 \equiv p_{n+2}) + \ldots + (p_{n+1} \equiv p_{n+2}).$$

The formulas can be also be written down in the form

$$\varphi_n = \oplus\{p_i \equiv p_j : 1 \leqslant i < j \leqslant n+1\}.$$

The set $A_l \cup \{\varphi_n\}$ is denoted by H_n. The completeness theorem for the system $\langle R_0, Sb(H_n)\rangle$ with respect to the algebra $\mathscr{G}_n$ will be formulated as follows:

Theorem 2.86. $Cn(R_0, Sb(H_n) \cup X) = \overrightarrow{\mathscr{G}_n}(X)$, *for each* $X \subseteq S_2$.

Proof. The algebra $\mathscr{G}_n$, for $n \geqslant 1$, is prime and contains exactly n elements. Then the formula $\oplus\{p_i \equiv p_j : 1 \leqslant i < j \leqslant n+1\}$ is valid in $\mathscr{G}_n$ and is not valid in $\mathscr{G}_{n+1}$, hence $\varphi_n \in E(\mathscr{G}_n) \setminus E(\mathscr{G}_{n+1})$. By Theorem 2.85 (i), (iv), we get

$$Cn(R_0, Sb(H_n)) = Cn(R_0, Sb(A_l) \cup Sb(\varphi_n)) = E(\mathscr{G}_n).$$

Obviously, the inclusion $Cn(R_0, Sb(H_n) \cup X) \subseteq \overrightarrow{\mathscr{G}_n}(X)$ holds for each $X \subseteq S_2$ (see 2.64 (ii)). On the other hand, it follows from the finiteness of $\overrightarrow{\mathscr{G}_n}$ (2.65) and from the deduction theorem Lemma 2.64 (ii) that $\overrightarrow{\mathscr{G}_n}(X) \subseteq Cn(R_0, Sb(H_n) \cup X)$. $\square$

Instead of the algebra $\mathscr{G}_n = \langle G_n, \dotrightarrow, \cup, \cap, -\rangle$, one could take the matrix $\mathfrak{G}_n = \langle\langle G_n, \dotrightarrow, \cup, \cap, -\rangle, \{1\}\rangle$. The completeness theorem for $\langle R_0, Sb(H_n)\rangle$ usually means adequacy with respect to the matrix $\mathfrak{G}_n$ that is as the equation

$$\overrightarrow{\mathfrak{G}_n}(X) = Cn(R_0, Sb(H_n) \cup X), \quad \text{for each } X \subseteq S_2$$

or as $E(\mathfrak{G}_n) = Cn(R_{0*}, H_n)$. Of course, $E(\mathscr{G}_n) = E(\mathfrak{G}_n)$ and, since $\{1\}$ is a filter in $\mathscr{G}_n$, one gets $\overrightarrow{\mathscr{G}_n}(X) \subseteq \overrightarrow{\mathfrak{G}_n}(X)$ for each X. The reverse inclusion can also be easily deduced. Namely, if $\alpha \notin \overrightarrow{\mathscr{G}_n}(X)$, then $h^v(X) \subseteq H$ and $h^v(\alpha) \notin H$ for some filter H in $\mathscr{G}_n$ and some valuation v. The quotient algebra $\mathscr{G}_n/H$ is isomorphic to some algebra $\mathscr{G}_m$ with $m \leqslant n$ and obviously $\alpha \notin \overrightarrow{\mathfrak{G}_m}(X)$. But $\mathfrak{G}_m$ is isomorphic to some submatrix of $\mathfrak{G}_n$. Thus $\overrightarrow{\mathfrak{G}_n}(X) \subseteq \overrightarrow{\mathfrak{G}_m}(X)$ and hence $\alpha \notin \overrightarrow{\mathfrak{G}_n}(X)$.

Using the completeness (adequacy) theorem we easily deduce

$$\mathrm{Der}(R_0, Sb(H_n)) = N(\mathscr{G}_n) = N(\mathfrak{G}_n).$$

Similar statements hold for the rules derivable in $\langle R_{0*}, H_n\rangle$.

Corollary 2.87. *For each* $X \subseteq S_2$,

$$Cn(R_{0*}, H_n \cup X) = \bigcap\{E(\mathscr{G}_m) : m \leqslant n \,\wedge\, X \subseteq E(\mathscr{G}_m)\}.$$

The above corollary follows directly from 2.85 and 2.86. We also get

$$Cn(R_{0*}, H_n \cup X) = E(\mathscr{G}_m) \quad \text{if} \quad m = \max\{i \leqslant n : X \subseteq E(\mathscr{G}_i)\}.$$

Thus, $\mathrm{Der}(R_{0*}, H_n) = \bigcap \{V(\mathscr{G}_m) : m \leqslant n\}$. It is worth noticing that instead of $\oplus \{p_i \equiv p_j : 1 \leqslant i < j \leqslant n+1\}$ one could take, as an axiom of Gödel–Heyting's logic, any formula φ_n fulfilling the condition $\varphi_n \in E(\mathscr{G}_n) \setminus E(\mathscr{G}_{n+1})$. Note that if $\varphi_n \in E(\mathscr{G}_n) \setminus E(\mathscr{G}_{n+1})$, then φ_n must contain (at least) $n+1$ variables.

Heyting's matrix $\mathfrak{G}_3$ was axiomatized by J. Łukasiewicz [65], 1938. The axiomatization of $\mathfrak{G}_n$ for $n > 3$ is due to I. Thomas [121], 1962. The linear propositional logic was defined, and the completeness theorem for this logic was proved by M.Dummett [16], 1959. In the proof of completeness theorem for $\langle R_{0*}, A_l \rangle$ (as well as for $\langle R_{0*}, H_n \rangle$) we used some ideas from Dunn [17], 1971. 1962).

Classical propositional logic

The system $\langle R_0, Sb(A_2) \rangle$ is formalized in $\mathscr{S}_2$ and FC_2 is the filter consequence connected with the system.

Corollary 2.88. *The set $FC_2(\emptyset)$ of all Boolean tautologies is the set of those formulas $\alpha \in S_2$ which in every Boolean algebra $\mathscr{B}$ fulfill the condition*

$$h^v(\alpha) = 1_B, \quad \text{for all } v: At \to B.$$

There exist, of course, Boolean tautologies which are not linear tautologies (which means that they are not Heyting tautologies, either). For instance, note that $((p \to q) \to p) \to p$ is not valid in $\mathscr{G}_3$ but is valid in each Boolean algebra.

Lemma 2.89. $A_2 \subseteq FC_2(\emptyset)$.

Theorem 2.90. $Cn(R_0, Sb(A_2) \cup X) = FC_2(X)$, *for each* $X \subseteq S_2$.

Proof. Inclusion $Cn(R_0, Sb(A_2) \cup X) \subseteq FC_2(X)$ for each $X \subseteq S_2$ follows from Lemma 2.89 and Corollary 2.73 (iii). The proof of the converse implication runs as follows. Let

$$\varphi \approx_X \psi \qquad \text{iff} \qquad (\varphi \equiv \psi) \in Cn(R_0, Sb(A_2) \cup X).$$

The relation is a congruence in the algebra of the language $\mathscr{S}_2$. In the quotient algebra $\mathscr{S}_2 / \approx_X$ we can define the order relation

$$[\varphi] \leqslant_X [\psi] \qquad \text{iff} \qquad (\varphi \to \psi) \in Cn(R_0, Sb(A_2) \cup X)$$

which is the order relation in the Boolean algebra $\mathscr{S}_2 / \approx_X$. Moreover,

$$[\varphi] = 1_X \qquad \text{iff} \qquad \varphi \in Cn(R_0, Sb(A_2) \cup X).$$

Now assume that $h: \mathscr{S}_2 \to \mathscr{S}_2 / \approx_X$ is the canonical homomorphism, i.e.,

$$h(\varphi) = [\varphi] \text{ for every } \varphi \in S_2.$$

Suppose that $\alpha \in FC_2(X)$, thus $h(\alpha) \in F(h(X))$ where F is the filter operation in $\mathscr{S}_2/\approx_X$. But we have

$$h(X) = \{[\beta] : \beta \in X\} = \{1_X\}.$$

So we get $[\alpha] = h(\alpha) \in F(h(X)) = \{1_X\}$, hence $[\alpha] = 1_X$, which means that $\alpha \in Cn(R_0, Sb(A_2) \cup X)$.   □

The above proof is carried out without the use of the deduction theorem and is based only on the construction of the Lindenbaum–Tarski algebra.

The above can be deduced from Corollary 2.77. Namely, we have to prove the inclusion $FC_2(\emptyset) \subseteq Cn(R_{0*}, A_2)$ only. Suppose $\alpha \notin Cn(R_0, Sb(A_2)) = FC_i(Sb(p+ \sim p))$. By the definition of FC_i and by 2.77 we get the existence of a Heyting algebra $\mathscr{A}$ such that $p+ \sim p \in E(\mathscr{A})$ and $\alpha \notin E(\mathscr{A})$. Since $p+ \sim p \in E(\mathscr{A})$, $\mathscr{A}$ is a Boolean algebra and hence $\alpha \notin FC_2(\emptyset)$.

By a similar argument, one gets the so-called Glivenko's theorem which states certain connections between the intuitionistic and classical propositional logics:

Corollary 2.91. $\alpha \in FC_2(X) \quad \Leftrightarrow \quad \sim\sim \alpha \in FC_i(X), \quad$ *for each* $\alpha, X \subseteq S_2$.

Now we will prove the completeness theorem for classical propositional logic with respect to the two-element Boolean algebra

$$\mathscr{B}_2 = \langle \{0,1\}, \overset{\rightarrow}{\rightarrow}, \cup, \cap, - \rangle.$$

Lemma 2.92. *If* $\mathscr{B}$ *is a non-degenerate Boolean algebra, then* $\overrightarrow{\mathscr{B}} = \overrightarrow{\mathscr{B}_2}$.

Proof. Grounds should be found only for $\overrightarrow{\mathscr{B}_2} \leqslant \overrightarrow{\mathscr{B}}$. Assume that $\alpha \notin \overrightarrow{\mathscr{B}}(X)$. Then, by Corollary 2.68, there exists a prime filter H such that $\alpha \notin \overrightarrow{\mathscr{B}/H}(X)$. It follows now from Corollary 1.22 that H is an ultrafilter and $\mathscr{B}/H$ contains exactly two elements. Since all two-element Boolean algebras are isomorphic, we conclude $\alpha \notin \overrightarrow{\mathscr{B}_2}(X)$.   □

The above lemma, founded on the basic theorems and the notions of the theory of Boolean algebras (cf. 1.20, 1.22, 2.62, 2.68, 2.67), is of course the crucial point in passing from a rather trivial type of Boolean completeness theorems to the non-trivial theorem on the $\mathscr{B}_2$-completeness of classical propositional logic. Lemma 2.92 is a generalization of some lemma from [86], 1974, which presents the very idea of 2.92 but is in less general form.

By Theorem 2.90 and Lemma 2.92 we get immediately

Corollary 2.93. $Cn(R_0, Sb(A_2) \cup X) = \overrightarrow{\mathscr{B}_2}(X),$ *for each* $X \subseteq S_2$.

So, we get decidability of the set of Boolean tautologies:

$$Cn(R_{0*}, A_2) = E(\mathscr{B}_2) = FC_2(\emptyset).$$

Moreover, The reader should notice close connections between

- the completeness theorem for the classical propositional logic with respect to the two-valued matrix $\mathfrak{M}_2$,

- the completeness of Gödel–Heyting's propositional logic $\langle R_0, Sb(H_2) \rangle$ in relation to the algebra $\mathcal{G}_2$ (or the matrix $\mathfrak{G}_2$, see 2.86),

- the completeness of Łukasiewicz's logic $\langle R_0, Sb(\text{Ł}_2) \rangle$ with respect to $\mathfrak{M}_2$.

Since any quotient algebra of $\mathcal{B}_2$ is either degenerate or is isomorphic with $\mathcal{B}_2$, we obtain on the basis of Corollaries 2.93 and 2.69:

Corollary 2.94. *For each $X \subseteq S_2$,*

$$Cn(R_{0*}, A_2 \cup X) = \begin{cases} E(\mathcal{B}_2) & \text{if } X \subseteq E(\mathcal{B}_2) \\ S_2 & \text{otherwise.} \end{cases}$$

From Corollaries 2.93 and 2.94 it follows that $\text{Der}(R_0, Sb(A_2)) = N(\mathcal{B}_2)$ and $\text{Der}(R_{0*}, A_2) = = V(\mathcal{B}_2)$. It should also be noticed that all above theorems concerning the two-element algebra $\mathcal{B}_2$ hold as well for each non-degenerate Boolean algebra $\mathcal{B}$.

Chapter 3

Completeness of propositional logics

The purpose of this chapter is to give a systematic treatment of the most important results concerning different notions of completeness for propositional logics. We consider, in Section 3.1, the notion of Γ-completeness and Γ-maximality and use them in the further development of the theory of Post-complete (Section 3.2) and structurally complete (Section 3.4) systems. Thus, Section 3.1 is rather technical; we search there for properties which Post-completeness, structural completeness, maximality and other similar notions have in common.

Sections 3.2–3.4 develop the basic results of this chapter. We discuss there the constructions, basic properties and several applications of complete systems. Extensions of a given propositional system to a complete one play the central role in Section 3.3. We try to determine propositional logics which can be extended to a Post-complete system only in a unique way. Some related concepts such as saturation and pseudo-completeness are discussed in Section 3.5. We give also some results concerning completeness of some concrete propositional logics. However, these results are limited since the main purpose of this chapter is to present only methods and ideas that have been used in this part of logic.

3.1 Generalized completeness

Let $\mathscr{S} = \langle S, F_1, \ldots, F_m \rangle$ be the algebra of a propositional language based on a countable set of propositional variables At and the connectives $F_1, \ldots, F_m$.

In the next definition the set $\Gamma \subseteq S$ will be used as a parameter; $\mathscr{R}_S$ is the set of all rules over S, the symbol $\text{Struct}(\Gamma)$ stands for the family of all Γ-structural rules (see Definition 1.54). Recall that a logic $\langle R, A \rangle$, where $R \subseteq \mathscr{R}_S$, $A \subseteq S$ is said to be Γ-invariant (in symbols $\langle R, A \rangle \in \text{Inv}(\Gamma)$) iff $R \subseteq \text{Struct}(\Gamma)$, $A = Sb_\Gamma(A)$ and that $\langle R, A \rangle$ is consistent, $\langle R, A \rangle \in \text{Cns}$, iff $Cn(R, A) \neq S$.

Definition 3.1 (a). Let $A \subseteq S$ and $R \subseteq \mathscr{R}_S$. Then

(i) $\langle R, A \rangle \in \Gamma\text{-Cpl} \Leftrightarrow \mathrm{Adm}(R, A) \cap \mathrm{Struct}(\Gamma) \subseteq \mathrm{Der}(R, A)$;

(ii) $\langle R, A \rangle \in \Gamma\text{-Max} \Leftrightarrow \big(\langle R, A \rangle \in \mathrm{Inv}(\Gamma)$ and there is no system $\langle R_1, A_1 \rangle \in \mathrm{Inv}(\Gamma) \cap \mathrm{Cns}$ such that $\langle R, A \rangle \prec \langle R_1, A_1 \rangle\big)$.

This definition can be expressed, as well, in the consequence operation formalism. Let Cn be a consequence operation over S, then we have:

Definition 3.1 (b).

(i) $Cn \in \Gamma\text{-CPL} \Leftrightarrow \mathrm{ADM}(Cn) \cap \mathrm{Struct}(\Gamma) \subseteq \mathrm{DER}(Cn)$;

(ii) $Cn \in \Gamma\text{-MAX} \Leftrightarrow Cn(\emptyset) = S$ or Cn is a maximal element in the family of all consistent Γ-structural consequence operations over S.

Note that the notion of Γ-completeness is defined for arbitrary propositional logics, whereas only Γ-invariant ones can be Γ-maximal.

Lemma 3.2. $\Gamma\text{-Max} \subseteq \Gamma\text{-Cpl} \cap \mathrm{Inv}(\Gamma)$.

Proof. It suffices to consider only the case when $\langle R, A \rangle \in \Gamma\text{-Max} \cap \mathrm{Cns}$. Let $R_1 = \mathrm{Adm}(R, A) \cap \mathrm{Struct}(\Gamma)$, then $\langle R, A \rangle \prec \langle R_1, A \rangle$ and $\langle R_1, A \rangle \in \mathrm{Cns}$. By our assumptions, we get $R_1 \subseteq \mathrm{Der}(R, A)$, i.e., $\langle R, A \rangle \in \Gamma\text{-Cpl}$. $\qquad\square$

We have $\Gamma\text{-Cpl} \cap \mathrm{Inv}(\Gamma) \neq \Gamma\text{-Max}$ for some $\Gamma \subseteq S$: the pure implicational Hilbert's propositional logic $\langle R_0, Sb(A_H^{\to}) \rangle$ belongs to the family $S\text{-Cpl}$ ([98], 1972) but, of course, not to $S\text{-Max}$. For the reader less familiar with such particular results concerning structural completeness we give a straightforward example:

Example. Let $\mathscr{S}_F = \langle S_F, F \rangle$ be the propositional language built up by means of one monadic connective F. Let $A = Sb(\{FFp\})$ and $R = \mathrm{Adm}(\emptyset, A) \cap \mathrm{Struct}$. We have $Cn(R, A) = Cn(\emptyset, A) = A$ and hence, $\mathrm{Adm}(R, A) = \mathrm{Adm}(\emptyset, A)$. Thus, $\langle R, A \rangle \in S\text{-Cpl}$. Let $r = \{\langle \varphi, \psi \rangle : \varphi \in A \ \wedge \ \psi \in Sb(\{Fp\})\}$. It is easy to see that $r \in \mathrm{Struct}$ and that $Fp \notin A = Cn(R, A)$. Thus $r \notin \mathrm{Der}(R, A)$ and hence $\langle R, A \rangle \prec \langle R \cup \{r\}, A \rangle$. Now we are going to show that $\langle R \cup \{r\}, A \rangle \in \mathrm{Cns}$. Of course, we have $Cn(R \cup \{r\}, A) = Cn(R, Sb(\{Fp\}))$. Suppose on the contrary that $p \in Cn(R, Sb(\{Fp\}))$. Take $e: At \to S$ such that $e(\gamma) = Fp$ for each $\gamma \in At$, we get $Fp = e(p) \in Cn\big(R, h^e(Sb(\{Fp\}))\big) \subseteq Cn\big(R, Sb(\{FFp\})\big) = Cn(R, A) = A$ which is impossible. This shows $\langle R \cup \{r\}, A \rangle \in \mathrm{Cns}$ and hence $\langle R, A \rangle \notin S\text{-Max}$.

The notions $\Gamma\text{-Max}$ and $\Gamma\text{-Cpl}$ can be defined by similar patterns.

Lemma 3.3. *If* $\langle R, A \rangle \in \mathrm{Inv}(\Gamma)$, *then*

(i) $\langle R, A \rangle \in \Gamma\text{-Max} \Leftrightarrow \big(\langle R \cup \{r\}, A \rangle \notin \mathrm{Cns}$, *for each* $r \in \mathrm{Struct}(\Gamma) \setminus \mathrm{Der}(R, A)\big)$;

(ii) $\langle R, A \rangle \in \Gamma\text{-Cpl} \Leftrightarrow \big(\langle R \cup \{r\}, A \rangle \notin \mathrm{Cns}$, *for each* $r \in \mathrm{Struct}(\Gamma) \cap \mathrm{Adm}(R, A) \setminus \mathrm{Der}(R, A)\big)$.

Easy proofs are omitted. The assumption $\langle R, A \rangle \in \mathrm{Inv}(\Gamma)$ is superfluous in the case of Lemma 3.3 (ii).

The next theorem reveals some connections between the rules derivable in $\langle R, A \rangle \in \Gamma\text{-Cpl}$ and rules valid in any $\langle R, A \rangle$-adequate matrix $\mathfrak{M}$.

Theorem 3.4. *If $r_* | \Gamma \in \mathrm{Adm}(R, A)$ and $\emptyset \neq Cn(R, A) \neq S$, then the following two conditions are equivalent:*

(i) $\mathrm{Der}(R, A) \cap \mathrm{Struct}(\Gamma) = V(\mathfrak{M}) \cap \mathrm{Struct}(\Gamma);$

(ii) $\langle R, A \rangle \in \Gamma\text{-Cpl} \;\wedge\; E(\mathfrak{M}) = Cn(R, A).$

Proof. (i)$\Rightarrow$(ii): Let $r_1 = \{ \langle \psi, \varphi \rangle : \psi \in Cn(R, A) \;\wedge\; \varphi \in E(\mathfrak{M}) \}$. From the assumptions it easily follows that $E(\mathfrak{M}) \neq \emptyset$. Then we get $r_1 \in V(\mathfrak{M}) \cap \mathrm{Struct}(\Gamma)$. Thus $r_1 \in \mathrm{Der}(R, A)$ and hence $E(\mathfrak{M}) \subseteq Cn(R, A)$.

Now, let us take $r_2 = \{ \langle \psi, \varphi \rangle : \psi \in E(\mathfrak{M}) \wedge \varphi \in Cn(R, A) \}$. Then we easily get $r_2 \in \mathrm{Der}(R, A) \cap \mathrm{Struct}(\Gamma)$ and $r_2 \in V(\mathfrak{M})$. Hence $Cn(R, A) \subseteq E(\mathfrak{M})$.

Assume that $r \in \mathrm{Adm}(R, A) \cap \mathrm{Struct}(\Gamma)$. Since $E(\mathfrak{M}) = Cn(R, A)$, then $r \in V(\mathfrak{M}) \cap \mathrm{Struct}(\Gamma)$, so $r \in \mathrm{Der}(R, A)$. Thus we get $\langle R, A \rangle \in \Gamma\text{-Cpl}$.

(i)$\Leftarrow$(ii): Assume that $Cn(R, A) = E(\mathfrak{M})$. We have $V(\mathfrak{M}) = \mathrm{Adm}(R, A)$ and hence $V(\mathfrak{M}) \cap \mathrm{Struct}(\Gamma) = \mathrm{Adm}(R, A) \cap \mathrm{Struct}(\Gamma) = \mathrm{Der}(R, A) \cap \mathrm{Struct}(\Gamma)$ by Γ-completeness of $\langle R, A \rangle$. $\qquad\square$

Corollary 3.5. *If $Cn(R, A) = E(\mathfrak{M})$, then*

$$\mathrm{Der}(R, A) \cap \mathrm{Struct}(\Gamma) = V(\mathfrak{M}) \cap \mathrm{Struct}(\Gamma) \quad \Leftrightarrow \quad \langle R, A \rangle \in \Gamma\text{-Cpl}.$$

The main conclusion we can infer from Theorem 3.4 is that, given any Γ-incomplete logic and a matrix $\mathfrak{M}$ adequate for this logic, the set of its derivable rules is not equal to the set of $\mathfrak{M}$-valid rules. This seems to be the very idea of incompleteness expressed in terms of derivability of inferential rules. From Theorem 3.4 we also get the following characterization of Γ-completeness by use of the rules normal in the Lindenbaum matrix $\mathfrak{M}^{R,A}$, see Corollary 2.43;

Corollary 3.6. *If $r_* \in \mathrm{Adm}(R, A)$, then*

$$\mathrm{Der}(R, A) \cap \mathrm{Struct}(\Gamma) = V(\mathfrak{M}^{R,A}) \cap \mathrm{Struct}(\Gamma) \quad \Leftrightarrow \quad \langle R, A \rangle \in \Gamma\text{-Cpl}.$$

Let us define now the '*big Γ-rule*' (see Definition 1.53). Note that $h^{e_n} \circ \ldots \circ h^{e_1}$ means the usual composition of the mappings involved.

Definition 3.7. For each $X \subseteq S$,

$$\langle \Pi, \alpha \rangle \in \Gamma\text{-}r_X \quad \Leftrightarrow \quad \big(h^{e_n} \circ \ldots \circ h^{e_1}(\Pi) \subseteq X \;\Rightarrow\; h^{e_n} \circ \ldots \circ h^{e_1}(\alpha) \in X,$$

$$\text{for each } n \geqslant 0 \text{ and each } e_1, \ldots, e_n \colon At \to \Gamma \big).$$

It should be noticed that the case $n = 0$ is not excluded in the above definition and hence, if $\langle \Pi, \alpha \rangle \in \Gamma\text{-}r_X$ and $\Pi \subseteq X$, then also $\alpha \in X$. Observe that for $\Gamma = S$ we have $S\text{-}r_X = r_X$ (see Definition 1.53).

Lemma 3.8. $\langle R, A \rangle \in \Gamma\text{-Cpl} \quad \Leftrightarrow \quad \Gamma\text{-}r_{Cn(R,A)} \in \text{Der}(R, A)$.

Proof. ($\Rightarrow$): Is is easy to see that $\Gamma\text{-}r_X \in \text{Struct}(\Gamma)$ for every $X \subseteq S$. Moreover $\Gamma\text{-}r_{Cn(R,A)} \in \text{Adm}(R, A)$.

($\Leftarrow$): Let us observe that for every rule $r \in \text{Adm}(R, A) \cap \text{Struct}(\Gamma)$ we have $r \subseteq \Gamma\text{-}r_{Cn(R,A)}$: if we assume $\langle \Pi, \alpha \rangle \in r$ and $e_1, \ldots, e_n : At \to \Gamma$, then we get $\langle h^{e_n} \ldots h^{e_1}(\Pi), h^{e_n} \ldots h^{e_1}(\alpha) \rangle \in r \in \text{Adm}(R, A)$. Thus, the derivability of $\Gamma\text{-}r_{Cn(R,A)}$ yields the derivability of all Γ-structural and admissible rules. $\square$

Careful examination of Definitions 3.1 leads to the conclusion that the notion of Γ-Cpl is defined in a manner essentially different from the notion of Γ-Max. Γ-completeness is defined only by internal properties of $\langle R, A \rangle$ expressed by the operations Adm and Der, whereas Γ-maximality can be treated as a property determined by the structure of all Γ-invariant logics. We can say then that Γ-Cpl is an internal property, whilst Γ-Max is an external property (of a propositional logic). However, Lemma 3.3 shows that the external property Γ-Max, of a Γ-invariant logic, can also be expressed by internal (local) properties of the logic. On the other hand, Γ-complete logics may possess some global properties referring to the structure of the set $\text{Inv}(\Gamma)$.

Theorem 3.9. *Let* $\langle R, A \rangle \in \text{Inv}(\Gamma)$. *Then*

$$\langle R, A \rangle \in \Gamma\text{-Cpl} \quad \Leftrightarrow \quad ([Cn(R, A) = Cn(R_1, A_1) \Rightarrow \langle R_1, A_1 \rangle \preccurlyeq \langle R, A \rangle],$$

$$\text{for every } \langle R_1, A_1 \rangle \in \text{Inv}(\Gamma)).$$

Proof. Let $\langle R, A \rangle \in \Gamma\text{-Cpl}$ and let $Cn(R, A) = Cn(R_1, A_1)$ for $\langle R_1, A_1 \rangle \in \text{Inv}(\Gamma)$. Then $R_1 \subseteq \text{Adm}(R_1, A_1) \cap \text{Struct}(\Gamma) = \text{Adm}(R, A) \cap \text{Struct}(\Gamma) \subseteq \text{Der}(R, A)$ and $A_1 \subseteq Cn(R, A)$, that is $\langle R_1, A_1 \rangle \preccurlyeq \langle R, A \rangle$. For the reverse implication: if $R_1 = \text{Adm}(R, A) \cap \text{Struct}(\Gamma)$, then $Cn(R, A) = Cn(R_1, A)$ (see Definition 1.45) and hence $\text{Adm}(R, A) \cap \text{Struct}(\Gamma) \subseteq \text{Der}(R, A)$, i.e., $\langle R, A \rangle \in \Gamma\text{-Cpl}$. $\square$

A system $\langle R, A \rangle$ is Γ-complete iff $\langle R, A \rangle$ is maximal in the family of all Γ-invariant logics with a fixed set of theorems. The assumption $\langle R, A \rangle \in \text{Inv}(\Gamma)$ (in the previous proposition) cannot be omitted. It can be only weakened: instead of it one can assume that $r_*|\Gamma \in \text{Adm}(R, A)$ or $Cn(R_2, A_2) = Cn(R, A)$ for some $\langle R_2, A_2 \rangle \in \text{Inv}(\Gamma)$.

There can be easily given somewhat different (but similar to 3.9) characterization of Γ-completeness:

Lemma 3.10. $\langle R, A \rangle \in \Gamma\text{-Cpl} \Leftrightarrow ([Cn(R, A) = Cn(R_1, A_1) \Rightarrow \langle R_1, A_1 \rangle \preccurlyeq \langle R, A \rangle],$

$$\text{for every } A_1 \subseteq S \text{ and } R_1 \subseteq \text{Struct}(\Gamma)).$$

Thus, Γ-Cpl can be treated as maximality in some family of propositional logics, in other words, as external property. We will prove now

Lemma 3.11. *Let* $\langle R, A \rangle \in \mathrm{Inv}(\Gamma)$. *Then*

(i) $\langle R, A \rangle \in \Gamma\text{-Max} \Leftrightarrow \big([\langle R \cup \{r\}, A \rangle \in \mathrm{Cns}$
$\Rightarrow \mathrm{Adm}(R \cup \{r\}, A) \cap \mathrm{Struct}(\Gamma) \subseteq \mathrm{Der}(R, A)]$, *for every* $r \in \mathscr{R}_S \big)$;

(ii) *if* $\langle R, A \rangle \in \Gamma\text{-Max}$, *then* $Cn(R, Sb_\Gamma\{\alpha\} \cup A) = S$ *for each* $\alpha \notin Cn(R, A)$.

Proof. (i): ($\Rightarrow$) Let us assume that $\langle R, A \rangle \in \Gamma\text{-Max}$ and $\langle R \cup \{r\}, A \rangle \in \mathrm{Cns}$. Let $R_1 = \mathrm{Adm}(R \cup \{r\}, A) \cap \mathrm{Struct}(\Gamma)$. Hence $\langle R_1, A \rangle \in \mathrm{Inv}(\Gamma)$, $R \subseteq R_1$ and $Cn(R_1, A) \subseteq Cn(R \cup \{r\}, A) \neq S$ and therefore $\langle R_1, A \rangle \in \mathrm{Cns}$. We have then $\Gamma\text{-Max} \ni \langle R, A \rangle \preccurlyeq \langle R_1, A_1 \rangle \in \mathrm{Cns} \cap \mathrm{Inv}(\Gamma)$, which means that $\langle R, A \rangle \approx \langle R_1, A \rangle$. This yields the inclusion $R_1 = \mathrm{Adm}(R \cup \{r\}, A) \cap \mathrm{Struct}(\Gamma) \subseteq \mathrm{Der}(R, A)$.

($\Leftarrow$): Let $r \in \mathrm{Struct}(\Gamma) \setminus \mathrm{Der}(R, A)$. If $\langle R \cup \{r\}, A \rangle \in \mathrm{Cns}$, then by the assumption we get $r \in \mathrm{Der}(R, A)$ which is impossible. Thus $\langle R \cup \{r\}, A \rangle \notin \mathrm{Cns}$ and, by Lemma 3.3 (i), we get $\langle R, A \rangle \in \Gamma\text{-Max}$.

(ii): Let $\langle R, A \rangle \in \Gamma\text{-Max}$ and let $\alpha \notin Cn(R, A)$. Defining $A_1 = A \cup Sb_\Gamma(\{\alpha\})$ we get $\langle R, A \rangle \prec \langle R, A_1 \rangle$ and of course $\langle R, A_1 \rangle \in \mathrm{Inv}(\Gamma)$. Thus, by Definition 3.1 (ii), we get $\langle R, A_1 \rangle \notin \mathrm{Cns}$, i.e., $Cn\big(R, A \cup Sb_\Gamma(\{\alpha\})\big) = S$. $\square$

From the two considered notions Γ-completeness is more generally defined. First, it is not restricted to the family of Γ-invariant logics. Second, Γ-maximality can be defined by use of Γ-completeness.

Theorem 3.12. *The following two conditions are equivalent:*

(i) $\langle R, A \rangle \in \Gamma\text{-Max}$;

(ii) $\langle R, A \rangle \in \Gamma\text{-Cpl} \cap \mathrm{Inv}(\Gamma)$ *and* $\langle R \cup \{r_* | \Gamma\}, A \rangle \in \emptyset\text{-Cpl}$.

Proof. (i)$\Rightarrow$(ii): Let us prove that $\mathrm{Adm}(R \cup \{r_* | \Gamma\}, A) \subseteq \mathrm{Der}(R \cup \{r_* | \Gamma\}, A)$ — see Lemma 3.2. Let $r \in \mathrm{Adm}(R \cup \{r_* | \Gamma\}, A) = \mathrm{Adm}(R, Sb_\Gamma(A))$ (see Theorem 1.58). Let us assume $\langle \Pi, \alpha \rangle \in r$ and consider the following two possibilities.

a. $\Pi \subseteq Cn\big(R, Sb_\Gamma(A)\big)$. Then $\alpha \in Cn\big(R, Sb_\Gamma(A)\big)$.

b. $\Pi \not\subseteq Cn\big(R, Sb_\Gamma(A)\big)$. Then by 3.11 (ii) we get $S = Cn\big(R, Sb_\Gamma(A \cup \Pi)\big)$.

Therefore $\alpha \in Cn(R \cup \{r_* | \Gamma, A \cup \Pi)$.

(ii)$\Rightarrow$(i): This follows immediately from 3.9 and 1.58. $\square$

Let us note that some of the above properties look better if formulated in consequence operations formalism. As an example, we reformulate Lemma 3.3. Let Cn_r be the consequence operation determined by the sole rule r (that is $Cn_r(X) = Cn(\{r\}, X)$) for each set $X \subseteq S$.

Lemma 3.3. *If* $Cn \in \mathrm{STRUCT}(\Gamma)$, *then*

(i) $Cn \in \Gamma\text{-MAX} \Leftrightarrow \forall_{r \in \mathrm{Struct}(\Gamma)}(r \notin \mathrm{DER}(Cn) \Rightarrow Cn \cup Cn_r \notin \mathrm{CNS})$;

(ii) $Cn \in \Gamma\text{-CPL} \Leftrightarrow \forall_{r \in \mathrm{Struct}(\Gamma)}(r \in \mathrm{ADM}(Cn) \setminus \mathrm{DER}(Cn) \Rightarrow Cn \cup Cn_r \notin \mathrm{CNS})$.

We can also reformulate Theorem 3.9 and Lemma 3.10; if $Cn \in \mathrm{STRUCT}(\Gamma)$, then

$$Cn \in \Gamma\text{-CPL} \quad \Leftrightarrow \quad \forall_{Cn_1 \in \mathrm{STRUCT}(\Gamma)} \ \big(Cn_1(\emptyset) = Cn(\emptyset) \Rightarrow Cn_1 \leqslant Cn\big),$$

$$Cn \in \Gamma\text{-CPL} \Leftrightarrow \forall_{A \subseteq S} \ \forall_{Cn_1 \in \mathrm{STRUCT}(\Gamma)} \ \big(Cn(\emptyset) = Cn_1(A) \Rightarrow Cn_1 \leqslant Cn\big).$$

Of course, not all propositional logics are complete, Γ-complete or Γ-maximal. As we know, there are many incomplete logics. Then, the problem of extending an incomplete logic to a complete one seems to be important. It is clear, however, that we can do such extensions in many various ways. We are interested in extensions which preserve the fundamental properties of initial logic (except of the incompleteness).

Theorem 3.13. *For every* $\langle R, A \rangle \in \mathrm{Cns}$ *there is a system* $\langle R_1, A_1 \rangle \in \Gamma\text{-Cpl}$ *such that* $\langle R, A \rangle \preccurlyeq \langle R_1, A_1 \rangle \in \mathrm{Cns}$.

Proof. Let $\langle R, A \rangle \in \mathrm{Cns}$ and take $R_1 = R \cup \mathrm{Adm}(R, A) \cap \mathrm{Struct}(\Gamma)$. We have $Cn(R, A) = Cn(R_1, A) \neq S$ and hence we get $\mathrm{Adm}(R_1, A) \cap \mathrm{Struct} \subseteq R_1$ as $\mathrm{Adm}(R_1, A) \cap \mathrm{Struct}(\Gamma) = \mathrm{Adm}(R, A) \cap \mathrm{Struct}(\Gamma)$. According to the definition of Γ-Cpl we have thus $\langle R, A \rangle \preccurlyeq \langle R_1, A \rangle$ and $\langle R_1, A \rangle \in \Gamma\text{-Cpl}$. $\square$

The possibility of extending any consistent logic $\langle R, A \rangle$ to a consistent and Γ-complete logic turns out to be trivial: it suffices to take $R_1 = \mathrm{Adm}(R, A)$ and $A_1 = A$. Note that the Γ-complete extension of $\langle R, A \rangle$ defined in the proof of Theorem 3.13 has, additionally, the following properties:

1. $Cn(R, A) = Cn(R_1, A)$,

2. $\langle R, A \rangle \in \Gamma\text{-Cpl} \Rightarrow \langle R, A \rangle \approx \langle R_1, A \rangle$,

3. $\langle R, A \rangle \in \mathrm{Inv}(\Gamma) \Rightarrow \langle R_1, A \rangle \in \mathrm{Inv}(\Gamma)$,

4. $\langle R, A \rangle \in \mathrm{Inv}(\Gamma) \wedge \langle R \cup \{r_* | \Gamma\}, A \rangle \in \emptyset\text{-Cpl} \Rightarrow \langle R_1, A \rangle \in \Gamma\text{-Max}$,

5. $\langle R_1, A \rangle \notin \Gamma\text{-Max}$, for some $\langle R, A \rangle \in \mathrm{Inv}(\Gamma)$ and some $\Gamma \subseteq S$.

Let us prove that Γ-maximality, in turn, can be achieved by any compact logic. We recall that $\langle R, A \rangle$ is compact, $\langle R, A \rangle \in \mathrm{Comp}$, iff for every $Y \subseteq S$ there is $X \in \mathrm{Fin}(Y)$ such that $Cn(R, A \cup X) = S$ whenever $Cn(R, A \cup Y) = S$.

Theorem 3.14. *If* $\langle R, A \rangle \in \mathrm{Inv}(\Gamma) \cap \mathrm{Cns}$ *and* $\langle R \cup \{r_* | \Gamma\}, A \rangle \in \mathrm{Comp}$, *then there is* $\langle R_1, A_1 \rangle \in \Gamma\text{-Max}$ *such that* $\langle R, A \rangle \preccurlyeq \langle R_1, A_1 \rangle \in \mathrm{Cns}$.

Proof. Consider the family $\{Y \subseteq S : S \neq Y = Cn\big(R, A \cup Sb_\Gamma(Y)\big)\}$. Any chain $\mathscr{L}$ in this family is bounded by $Cn(R, A \cup \bigcup \mathscr{L})$ (the system $\langle R \cup \{r_* | \Gamma\}, A \rangle$ is compact). By Zorn's lemma there exists in the considered family a maximal element. Thus $A \subseteq A_1 = Sb_\Gamma(A_1)$ and $\langle R \cup \{r_* | \Gamma\}, A_1 \rangle \in \emptyset\text{-Cpl} \cap \mathrm{Cns}$ for some $A_1 \subseteq S$. It follows now from 3.13 (see comments on that theorem) that there is $R_1 \subseteq \mathrm{Struct}(\Gamma)$ such that $R \subseteq R_1$ and $\langle R_1, A_1 \rangle \in \Gamma\text{-Max} \cap \mathrm{Cpl}$. $\square$

In the proof we have used Zorn's lemma. Let us remark that 3.14 can also be proved effectively if we use the fact that S is countable. Note, however, that the assumption $\langle R \cup \{r_* | \Gamma\}, A \rangle \in$ Comp cannot be omitted:

Example. Let $\mathscr{S} = \langle S_{F,G}, F, G \rangle$ where F, G are unary connective. Define $F^n p$ by induction: $F^0 p = p$, $F^{k+1} p = F F^k p$ and let $\alpha_n = F^n G p$. Let $e_n \colon At \to S$ be the mapping such that $e_n(\gamma) = F^n p$ for each $\gamma \in At$ and $n \geq 1$. We have

$$(\star) \qquad\qquad Sb(\{\alpha_n\}) \cap Sb(\{\alpha_m\}) = \emptyset, \qquad \text{for } n \neq m.$$

Denote $A = Sb(\{h^{e_n}(\alpha_n) : n \geq 1\})$ and consider the following structural rules:

$$r_1 = \{\langle \phi, \psi \rangle : \phi = h^e(\alpha_{i+1}) \wedge \psi = h^e(\alpha_i) \text{ for some } i \geq 1, e \colon At \to S\},$$
$$r_2 = \{\langle \phi, \psi \rangle : \phi \in Sb(\{Gp\}) \wedge \psi \in S\},$$
$$r_3 = \{\langle \Pi, \psi \rangle : \Pi = h^e(\{\alpha_i : i \geq 1\}) \wedge \psi \in S \text{ for some } e \colon At \to S\}.$$

Let $R = \{r_1, r_2, r_3\}$. We have, of course, $\langle R, A \rangle \in$ Inv and now we will show that $Cn(R, A) = Sb(\{h^{e_n}(\alpha_n) : n \geq 1\})$. The inclusion $(\supseteq)$ is obvious by the definition of A. Then, it suffices to prove that the set $Sb(\{h^{e_n}(\alpha_n) : n \geq 1\})$ is closed under the rules r_1, r_2, r_3. Assume that $h^e(\alpha_{i+1}) \in Sb(\{h^{e_n}(\alpha_n) : n \geq 1\})$, hence by $(\star)$ we get $h^e(\alpha_{i+1}) \in Sb(\{h^{e_{i+1}}(\alpha_{i+1})\})$. Thus $e(p) = F^{i+1}\alpha$ for some $\alpha \in S$. This yields, however, that $h^e(\alpha_i) = F^i G F^{i+1}\alpha \in Sb(\{h^{e_i}(\alpha_i)\})$ which means that the considered set is closed under the rule r_1; the fact that this set is closed also under the rule r_2 is obvious.

Now assume that $h^e(\bigcup\{\alpha_i : t \geq 1\}) \subseteq Sb(\{h^{e_n}(\alpha_n) : n \geq 1\})$ and let $i = l(e(p)) + 1$. By $(\star)$ we have $h^e(\alpha_i) \in Sb(\{h^{e_i}(\alpha_i)\})$ and $l(h^e(\alpha_i)) < l(h^{e_i}(\alpha_i))$ — which is impossible. Thus, the set $l(h^e(\alpha_i))$ is closed under the rule r_3.

Then $\langle R, A \rangle \in$ Cns. Let now $\langle R, A \rangle \preccurlyeq \langle R_1, A_1 \rangle \in$ Cns $\cap$ Inv. We will show

$$(\star\star) \qquad\qquad Cn(R_1, A_1 \cup \{\alpha_i\}) \neq S \quad \text{for each } i \geq 1.$$

Let $Gp \in Cn(R_1, A_1 \cup \{\alpha_i\})$, hence $Ge_i(p) = h^{e_i}(Gp) \in Cn(R_1, A_1 \cup \{h^{e_i}(\alpha_i)\}) \subseteq \subseteq Cn(R_1, A_1 \cup Cn(R, A)) \subseteq Cn(R_1, A_1)$. Thus $S = Cn(R, Ge_i(p)) \subseteq Cn(R_1, A_1)$, which is impossible. Then we get $Gp \notin Cn(R_1, A_1 \cup \{\alpha_i\})$. Moreover, we have $Cn(R_1, A_1 \cup \{\alpha_i : i \geq 1\}) \supseteq Cn(R, A \cup \{\alpha_i : i \geq 1\}) = S$ which means that $\alpha_{n_0} \notin Cn(R_1, A_1)$ for some $n_0 \geq 1$. Consider the rule r defined as follows:

$$r = \{\langle \phi, \psi \rangle : \phi \in S \wedge \psi \in Sb(\{\alpha_{n_0}\})\} \in \text{Struct}.$$

We have $Cn(R_1 \cup \{r\}, A_1) = Cn(R_1, A_1 \cup Sb(\{\alpha_{n_0}\})) \neq S$. Hence $\langle R_1 \cup \{r\}, A_1 \rangle \in$ Cns and $\langle R_1, A_1 \rangle \prec \langle R_1 \cup \{r\}, A_1 \rangle$. Thereby $\langle R_1, A_1, \rangle \notin S$-Max.

3.2 Post-completeness

Let us comment on Definition 3.1. If $\Gamma = \emptyset$, then $\text{Inv}(\Gamma)$ is the set of all propositional logics because $r_* | \emptyset$ is the empty rule. Instead of $\emptyset$-Max (see 3.1 (ii)) we can simply write Max, since any $\emptyset$-maximal logic (if inconsistent) is a maximal element in the (pre)ordered set of all consistent propositional logics.

Compare now the notions of Max, $\emptyset$-Cpl with the notion of Post-completeness, Cpl. Let us recall:

$$\langle R, A \rangle \in \text{Cpl} \Leftrightarrow \left(Cn(R, A \cup \{\alpha\}) = S, \text{ for each } \alpha \notin Cn(R, A) \right);$$

$$\langle R, A \rangle \in \emptyset\text{-Cpl} \Leftrightarrow \text{Adm}(R, A) \subseteq \text{Der}(R, A);$$

$$\langle R, A \rangle \in \text{Max} \Leftrightarrow \neg \left(\langle R, A \rangle \prec \langle R_1, A_1 \rangle \right), \text{ for each } \langle R_1, A_1 \rangle \in \text{Cns}.$$

We have, of course;

Theorem 3.15. $\emptyset$-Cpl $=$ Cpl $=$ Max.

The above three definitions determine then the same notion, namely that of Post-completeness. Moreover, we get

Corollary 3.16. $\langle R, A \rangle \in \text{Cpl} \Leftrightarrow \left(\langle R \cup \{r\}, A \rangle \notin \text{Cns}, \text{ for each } r \notin \text{Der}(R, A) \right).$

This follows from Theorem 3.3 (i), whereas Theorem 3.4 enables us to express simultaneously Post–completeness and matrix adequacy (see [84], 1973):

Theorem 3.17. *Let* $\mathfrak{M}$ *be a logical matrix and let* $\emptyset \neq Cn(R, A) \neq S$. *Then*

(i) $\text{Der}(R, A) = V(\mathfrak{M})$ $\Leftrightarrow$ $E(\mathfrak{M}) = Cn(R, A) \wedge \langle R, A \rangle \in \text{Cpl};$

(ii)   *If* $Cn(R, A) = E(\mathfrak{M})$, *then* $\text{Der}(R, A) = V(\mathfrak{M}) \Leftrightarrow \langle R, A \rangle \in \text{Cpl}.$

Thus, all $\mathfrak{M}$-valid rules, where $\mathfrak{M}$ is an $\langle R, A \rangle$-adequate matrix, are derivable in the Post-complete logic $\langle R, A \rangle$. That is to say, for such logics, derivability of a rule is equivalent to its $\mathfrak{M}$-validity. Since the Lindenbaum matrix $\mathfrak{M}^{R,A}$ is adequate for $\langle R, A \rangle$ (if the substitution rule is admissible) we get by 3.6

Corollary 3.18. *If* $r_* \in \text{Adm}(R, A)$, *then* $\text{Der}(R, A) = V(\mathfrak{M}^{R,A}) \Leftrightarrow \langle R, A \rangle \in \text{Cpl}.$

By Theorem 3.13 we get also

Corollary 3.19. *For every* $\langle R, A \rangle \in \text{Cns}$ *there is a system* $\langle R_1, A_1 \rangle \in \text{Cns} \cap \text{Cpl}$ *such that* $\langle R, A \rangle \preccurlyeq \langle R_1, A_1 \rangle.$

Every logic can be then extended to a Post–complete one — we obtained this extension by putting $R_1 = \text{Adm}(R, A)$. Thus, the logic $\langle R_1, A_1 \rangle$ satisfies the following conditions (besides those stated in 3.19):

1. $Cn(R, A) = Cn(R_1, A_1);$

2. $\langle R, A \rangle \in \text{Cpl}$ $\Rightarrow$ $\langle R, A \rangle \approx \langle R_1, A_1 \rangle.$

Complete extensions can be achieved in various ways. We present below a proof of Lindenbaum's theorem in which another construction of a complete extension is exposed.

Theorem 3.20. *If $\langle R, A \rangle \in$ Comp and $Cn(R, A \cup X) \neq S$, then there is a set $Y \subseteq S$ such that*

(i) $Cn(R, A \cup X) \subseteq Cn(R, A \cup Y) \neq S$;

(ii) $Cn(R, A \cup Y) = Y$;

(iii) $Cn(R, A \cup Y \cup \{\alpha\}) = S$, *for every* $\alpha \notin Y$.

Proof. Let us assume that $Cn(R, A \cup X) \neq S$. Since the set S is countable, there exists an infinite enumeration of all formulas from S:

$$\alpha_0, \alpha_1, \alpha_2, \ldots$$

Define now an infinite sequence of sets $X_0, X_1, X_2, \ldots,$

a. $X_0 = Cn(R, A \cup X)$

b., $X_{k+1} = \begin{cases} X_k & \text{iff } Cn(R, A \cup X_k \cup \{\alpha_k\}) = S \\ Cn(R, A \cup X_k \cup \{\alpha_k\}) & \text{iff } Cn(R, A \cup X_k \cup \{\alpha_k\}) \neq S. \end{cases}$

Note that $X_k \subseteq X_{k+1}$ and $X_k = Cn(R, A \cup X_k) \neq S$. Let $Y = \bigcup\{X_k : k \in \mathbb{N}\}$. Thus $X_0 \subseteq Y$. Assume $Cn(R, A \cup Y) = S$. Then there exist $\beta_1, \ldots, \beta_n \in Y$ such that $Cn(R, A \cup \{\beta_1, \ldots, \beta_n\}) = S$ and $\beta_1, \ldots, \beta_n \in X_i$ for some i. Thus, $X_i = Cn(R, A \cup X_i) = S$ which is impossible. The clause (i) is then proved. Suppose $\alpha \notin Y$. We have $\alpha = \alpha_i$ for some i, then $\alpha_i \notin X_j$, for each j. Let $\alpha_i \in Cn(R, A \cup Y) = Cn(R, A \cup \bigcup\{X_i : i \in \mathbb{N}\})$. Since $\{X_n : n \in \mathbb{N}\}$ is a chain of sets then $\alpha_i \in \bigcup\{Cn(R, A \cup X_i) : i \in \mathbb{N}\}$ which yields that $\alpha_i \in Cn(R, A \cup X_s)$ for some s and thus we get to a contradiction. Then (ii) is proved. Suppose that $\alpha \notin Y$ and $Cn(R, A \cup Y \cup \{\alpha\}) \neq S$. Then $\alpha = \alpha_i$ for some i and we get $X_{i+1} = Cn(R, A \cup X_i \cup \{\alpha_i\}) \neq S$, hence $\alpha_i \in X_{i+1}$ which is impossible. Then (iii) is also proved. $\square$

In the above proof we have assumed that S is countable. Theorem 3.20 can also be shown for uncountable languages; however, the Axiom of Choice is required. Notice that the system $\langle R, Y \rangle$, where Y is the superset constructed in the proof of 3.20, is a Post-complete oversystem of the logic $\langle R, X \rangle$, that is, $\langle R, X \rangle \preccurlyeq \langle R, Y \rangle \in$ Cns $\cap$ Cpl. Then, it is clear that the extension of a consistent logic $\langle R, X \rangle$ to a consistent, Post-complete one can be done in at least two ways: by extending the set R of rules or by extending the set X of axioms. The second method leads to the notion of the so-called *Lindenbaum supersets* of the set $Cn(R, X)$; the family of those supersets is defined as follows:

Definition 3.21.

$$\mathscr{L}(R, X) = \{Y : X \subseteq Y = Cn(R, Y) \neq S \wedge Cn(R, Y \cup \{\alpha\}) = S \text{ for each } \alpha \notin Y\}.$$

Considering the family of R-closed supersets of a given set $Cn(R, X)$ (a set $Y \subseteq S$ is R-closed iff $Cn(R, Y) = Y$) we are interested in whether there exists a consistent, maximal element in this family. The positive answer to this question was given by A. Lindenbaum in 1930 (see Tarski's paper [118], 1930) and has been known, since then, as Lindenbaum's theorem on maximal supersets (Theorem 3.20 or $\mathscr{L}(R, X) \neq \emptyset$, see 3.21). This result plays an important role in foundational studies. The algebraic counterpart of this theorem, proved in lattice theory by Stone in 1934 as the ultrafilter theorem, is an important proof-theoretical tool in contemporary mathematics. At the end of these comments we will write down Lindenbaum's theorem in the consequence operation formalism

Theorem 3.22. *If $Cn \in COMP$ and $Cn(X) \neq S$, then $\mathscr{L}(Cn, X) \neq \emptyset$, where $\mathscr{L}(Cn, X) = \{Y \subseteq S : X \subseteq Y = Cn(Y) \wedge Cn(Y \cup \{\alpha\}) = S$ for each $\alpha \notin Y\}$.*

It is clear that for every $Y \in \mathscr{L}(R, X)$ the system $\langle R, Y \rangle$ is uniquely determined. One can say that $\langle R, Y \rangle$ is a *Lindenbaum oversystem* of $\langle R, X \rangle$; certainly $\langle R, Y \rangle \in \text{Cns} \cap \text{Cpl}$. Then we will use both names: Lindenbaum superset Y and Lindenbaum oversystem $\langle R, Y \rangle$, if $Y \in \mathscr{L}(R, X)$.

There are very many Post-incomplete systems but it is intuitively clear that the incompleteness of some of them is, in a sense, greater (or smaller) than the incompleteness of other systems. Thus, it it worth introducing a notion which may be a 'measure of incompleteness' for propositional logics. Let us define the global *degree of incompleteness* of $\langle R, A \rangle$ as the cardinality of the family

$$\{\text{Der}(R \cup R', A \cup A') : A' \subseteq S \wedge R' \subseteq \mathscr{R}_S\}.$$

This is a very general notion; if we restrict this definition by the condition $\langle R', A' \rangle \in$ Inv, which means that $R' \subseteq$ Struct and $A' = Sb(A')$, then we get the *degree of maximality* of a given system $\langle R, A \rangle \in$ Inv.

This is the counterpart of the notion introduced by R.Wójcicki [143], 1974, in the consequence operation formalism. Namely, given a structural consequence Cn, the cardinality of the set $\{Cn' \in \text{Struct} : Cn \leqslant Cn'\}$ is said to be the degree of maximality of Cn. On the other hand, for $R' = \emptyset$ and for fixed R and A we get the notion of *degree of completeness* as introduced by A.Tarski ([118], 1930):

Definition 3.23. $dc(\langle R, A \rangle) = Nc\{Cn(R, A \cup X) : X \subseteq S\}$.

The latter notion was the first, from the chronological point of view, 'measure of incompleteness' in logic. (NB. We preserve the original name for $dc(\langle R, A \rangle)$, that is 'degree of completeness', though 'degree of incompleteness' would be appropriate.) Definition 3.23 can also be formulated in the consequence operation formalism: $dc(Cn)$ is then the cardinality of the set $\{Cn(X) : X \subseteq S\}$. The notion of global degree of incompleteness and the one introduced by Tarski, though related, are different: for the system $\langle R_{0*}, \text{Ł}_3 \rangle$ of Łukasiewicz the degree of completeness is equal to 3, but the global degree of incompleteness is equal to 4. Note that $dc(\langle R, A \rangle) \leqslant 2$ iff $\langle R, A \rangle \in$ Cpl; the same equivalence holds for the global degree of incompleteness. Moreover, let us notice that $dc(\langle R, A \rangle)$ for any invariant logic $\langle R, A \rangle$ is usually an infinite cardinal number.

We draw up the following two assertions (see [141], 1972 and [123], 1973), which follow directly from Theorem 2.38.

Corollary 3.24. *If $\mathfrak{M}$ is any finite matrix, then $dc(\overrightarrow{\mathfrak{M}} \circ Sb) < \aleph_0$.*

Corollary 3.25. *If $\mathfrak{M}$ is any functionally complete matrix, then $\overrightarrow{\mathfrak{M}} \circ Sb \in \mathrm{CPL}$, i.e., $dc(\overrightarrow{\mathfrak{M}} \circ Sb) \leqslant 2$.*

For finitistic logics (which are not compact) one can prove the so-called Lindenbaum–Asser theorem ([1], 1959, see also Łoś [61], 1951):

Theorem 3.26. *If Cn_{RA} is finitistic and if $\alpha \notin Cn(R, A \cup X)$, then there exists a set $Y \subseteq S$ such that*

a. $Cn(R, A \cup X) \subseteq Y = Cn(R, A \cup Y) \land \alpha \notin Y$;

b. $\alpha \in Cn(R, A \cup Y \cup \{\beta\})$, *for every* $\beta \notin Cn(R, A \cup Y)$.

The proof of this theorem is similar to the proof of Theorem 3.20, and is omitted here. Theorem 3.26 shows that in the family of all oversets of the set $Cn(R, A)$ we can define relative maximality, in other words, maximality with respect to non-deducibility of a given formula:

Definition 3.27.

$$\mathscr{L}^\alpha(R, A) = \{Y : X \subseteq Cn(R, Y) = Y \not\ni \alpha \in Cn(R, Y \cup \{\beta\}) \text{ for each } \beta \notin Y\}.$$

The notion of a *relative Lindenbaum superset* (see Definition 3.27) is weaker than the notion of Lindenbaum superset (see Definition 3.21), for: if we have $\alpha \notin Y \in \mathscr{L}(R, A \cup X)$, then $Y \in \mathscr{L}^\alpha(R, A \cup X)$ but not conversely. Besides, Theorem 3.26 can also be written down in the following form:

Theorem 3.26 (fin). *For every $\alpha \notin Cn(R, A)$ we have $\mathscr{L}^\alpha(R, A) \neq \emptyset$.*

It follows from the example on page 97 that the finiteness of a given consequence (logic) cannot be omitted here. Similarly as in the case of the family $\mathscr{L}(R, X)$; for every $Y \in \mathscr{L}^\alpha(R, X)$, the system $\langle R, Y \rangle$ is said to be a relative Lindenbaum oversystem of the system $\langle R, X \rangle$, and so we will use the name 'relative Lindenbaum superset Y' as well, as the name 'relative oversystem'.

Let us apply the introduced notions to oversystems of the classical propositional logic. Let $\mathscr{S}_2 = \langle S_2, \rightarrow, \cdot, +, \sim \rangle$ be the standard language (algebra of language) of propositional logics — most of the theorems proved here hold, however, also for sublanguages of $\mathscr{S}_2$. Let us recall that Cn_2 denotes the consequence operation generated by the invariant version of the classical logic $\langle R_0, Sb(A_2) \rangle$, i.e.,

$$Cn_2(X) = Cn(R_0, Sb(A_2) \cup X), \quad \text{for each } X \subseteq S_2.$$

One of the most important theorems on the classical propositional logic is the well-known deduction theorem which implies Tarski's lemma on consistency (see Corollary 1.67 (i),(iv) and (v)). Let us recall these properties.

Lemma 3.28. *For each $X \subseteq S_2$ and each $\alpha, \beta \in S_2$*

(i) $\alpha \in Cn_2(X \cup \{\beta\}) \quad \Leftrightarrow \quad (\beta \to \alpha) \in Cn_2(X)$;

(ii) $Cn_2(X \cup \{\sim \alpha\}) \neq S_2 \quad \Leftrightarrow \quad \alpha \notin Cn_2(X)$;

(iii) $Cn_2(X \cup \{\alpha\}) \neq S_2 \quad \Leftrightarrow \quad \sim \alpha \notin Cn_2(X)$.

Using the above properties of classical connectives, we immediately obtain the following characterization of consistent and complete extensions of classical (invariant) logic $\langle R_0, Sb(A_2) \cup X \rangle$ (or the structural consequence Cn_2):

Corollary 3.29. *For each $X \subseteq S_2$:*

(i) $\langle R_0, Sb(A_2) \cup X \rangle \in \mathrm{Cns} \Leftrightarrow \left(\neg(\alpha, \sim \alpha \in Cn_2(X),\ \text{for some } \alpha \in S_2)\right)$;

(ii) $\langle R_0, Sb(A_2) \cup X \rangle \in \mathrm{Cpl} \Leftrightarrow ((\alpha \in Cn_2(X) \vee \sim \alpha \in Cn_2(X),\ \text{for each } \alpha \in S_2))$.

Moreover, it easily follows from 3.29 that every relative Lindenbaum superset of $Cn(R_0, Sb(A_2))$ is also a Lindenbaum superset:

Corollary 3.30. $\mathscr{L}^\alpha(R_0, Sb(A_2)) \subseteq \mathscr{L}(R_0, Sb(A_2))$, *for each $\alpha \in S_2$.*

By 3.30 and 3.26 we get immediately the so-called Lindenbaum–Łoś's theorem (see [61], 1951) on complete extensions of the classical logic:

Theorem 3.31. *If $\alpha \notin Cn_2(X)$, then there is a set $Y \subseteq S_2$ such that*

(i) $Cn_2(X) \subseteq Y = Cn_2(Y) \not\ni \alpha$;

(ii) $Cn_2(Y \cup \{\beta\}) = S_2$, *for every $\beta \notin Y$.*

Theorem 3.31 can be written down in the following form:

Corollary 3.32. *For each $X \subseteq S_2$ and each $\alpha \in S_2$:*

(i) *If $\alpha \notin Cn_2(X)$, then there is $Y \in \mathscr{L}(R_0, Sb(A_2))$ such that $X \subseteq Y \not\ni \alpha$;*

(ii) $Cn_2(X) = \bigcap \{Y \subseteq S_2 : X \subseteq Y \in \mathscr{L}(R_0, Sb(A_2))\}$.

Corollary 3.32 shows that the logic $\langle R_0, Sb(A_2) \rangle$ is determined by all its Lindenbaum supersets. Now, we can state the filter-property of Lindenbaum oversets of the set of classical tautologies ([120], 1956) which is in fact an immediate corollary of Corollary 1.67.

Lemma 3.33. *For every $Y \in \mathscr{L}(R_0, Sb(A_2))$ we have*

(i) $\alpha \to \beta \in Y \quad \Leftrightarrow \quad (\alpha \in Y \Rightarrow \beta \in Y)$;

(ii) $\alpha + \beta \in Y \quad \Leftrightarrow \quad (\alpha \in Y \vee \beta \in Y)$;

(iii) $\alpha \cdot \beta \in Y \quad \Leftrightarrow \quad (\alpha \in Y \wedge \beta \in Y)$;

(iv) $\sim \alpha \in Y \quad \Leftrightarrow \quad \alpha \notin Y$.

We have proved, in Chapter 2, the completeness of the classical propositional logic with respect to the classical two-valued matrix $\mathfrak{M}_2$ (or the two-element Boolean algebra $\mathscr{B}_2$), see Corollary 2.93. Now, the above theorems concerning completeness enable us to give another proof of this result. We have

Lemma 3.34. $A_2 \subseteq E(\mathfrak{M}_2)$ *and* $r_0 \in N(\mathfrak{M}_2)$.

The notion of satisfiability in the matrix $\mathfrak{M}_2$ has been defined by the equivalence (see Definition 2.35)

$$X \in \mathrm{Sat}(\mathfrak{M}_2) \quad \Leftrightarrow \quad (h^v(X) \subseteq \{1\}, \text{ for some } v \colon At \to \{0,1\}).$$

One of the most important results of a general nature pertaining to the classical logic is the so-called Gödel–Malcev propositional theorem:

Theorem 3.35. $Cn_2(X) \neq S_2 \quad \Leftrightarrow \quad X \in \mathrm{Sat}(\mathfrak{M}_2)$, *for each* $X \subseteq S_2$.

Proof. ($\Rightarrow$): By the assumption and Theorem 3.31 we conclude that there is a set $Y \in \mathscr{L}(R_0, Sb(A_2) \cup X)$. Define the valuation $v \colon At \to \{0,1\}$,

$$v(\gamma) = 1 \quad \Leftrightarrow \quad \gamma \in Y, \quad \text{for every } \gamma \in At.$$

Then by Lemma 3.33, we have $h^v(\alpha) = 1 \Leftrightarrow \alpha \in Y$, for every $\alpha \in S$, which yields the double inclusion $h^v(X) \subseteq h^v(Y) \subseteq \{1\}$.

($\Leftarrow$): Assume the right side of 3.35. Using Lemma 3.34 we prove easily that $h^v(Cn(R_0, Sb(A_2) \cup X)) \subseteq \{1\}$ and $h^v(p \cdot \sim p) = 0$ for some $v \colon At \to \{0,1\}$. Then we get $Cn(R_0, Sb(A_2) \cup X) \neq S_2$. $\qquad\square$

An immediate consequence of 3.35 (and finiteness of $\langle R_0, Sb(A_2) \rangle$) is the compactness theorem for the notion of satisfiability:

Theorem 3.36. *For every* $X \subseteq S_2$,

$$X \in \mathrm{Sat}(\mathfrak{M}_2) \quad \Leftrightarrow \quad (Y \in \mathrm{Sat}(\mathfrak{M}), \text{ for each } Y \in \mathrm{Fin}(X)).$$

By Gödel–Malcev's propositional theorem 3.35 and Tarski's lemma 3.28(ii), we receive the completeness theorem for classical logic with respect to the two-valued matrix $\mathfrak{M}_2$. We formulate this theorem in a stronger version than usual; in the traditional terminology 3.37 (i) is called Post's theorem on completeness of the classical logic.

Theorem 3.37. (i) $Cn(R_0, Sb(A_2)) = Cn(R_{0*}, A_2) = E(\mathfrak{M}_2)$;

(ii) $\mathrm{Der}(R_0, Sb(A_2)) = N(\mathfrak{M}_2)$;

(iii) $\mathrm{Der}(R_{0*}, A_2) = V(\mathfrak{M}_2) = \mathrm{Adm}(R_0, Sb(A_2)) = \mathrm{Adm}(R_{0*}, A_2)$.

Proof. (ii): By 3.34, we have $\mathrm{Der}\big(R_0, Sb(A_2)\big) \subseteq N(\mathfrak{M}_2)$. Let $\langle \Pi, \alpha \rangle \in r \in N(\mathfrak{M}_2)$; if $\alpha \notin Cn\big(R_0, Ab(A_2) \cup \Pi\big)$, then $Cn\big(R_0, Sb(A_2) \cup \Pi \cup \{\sim \alpha\}\big) \neq S_2$ by 3.28. From 3.35 we infer that there exists a valuation $v \colon At \to \{0,1\}$ such that $h^v(\Pi) \subseteq \{1\}$ and $h^v(\alpha) = 0$; thus $r \notin N(\mathfrak{M}_2)$, which contradicts our assumptions.

(i): It follows from (ii) and Lemma 2.48.

(iii): By 3.34 we get $\mathrm{Der}(R_{0*}, A_2) \subseteq V(\mathfrak{M}_2)$. Let $r \in V(\mathfrak{M}_2) \setminus \mathrm{Der}(R_{0*}, A_2)$, we have then $\alpha \notin Cn(r_{0*}, A_2 \cup \Pi)$ for some $\langle \Pi, \alpha \rangle \in r$ and hence $\alpha \notin Cn_2\big(Sb(\Pi)\big)$. By 3.28, we get $Cn_2\big(Sb(\Pi) \cup \{\sim \alpha\}\big) \neq S_2$. Hence, by 3.35, there is $v \colon At \to \{0,1\}$ such that $h^v\big(Sb(\Pi)\big) \subseteq \{1\}$ and $h^v(\alpha) = 0$. Since $r \in V(\mathfrak{M}_2)$, we have $\beta \notin E(\mathfrak{M}_2)$ for some $\beta \in \Pi$. By (i), there is $v_1 \colon At \to \{0,1\}$ such that $h^{v_1}(\beta) = 0$. Hence, if we define a mapping $e \colon At \to S$ as follows

$$
e(\eta) = \begin{cases} p \to p & \text{iff } v_1(\eta) = 1 \\ \sim (p \to p) & \text{iff } v_1(\eta) = 0, \ \text{for } \eta \in At, \end{cases}
$$

then we get $\sim h^e(\beta) \in E(\mathfrak{M}_2)$. Thus $h^v h^e(\beta) = 0$ and $h^v\big(Sb(\Pi)\big) \subseteq \{1\}$, but $\beta \in \Pi$, then $h^v h^e(\beta) \in h^v\big(Sb(\Pi)\big)$, which is impossible. $\qquad\square$

Corollary 3.38. $Cn\big(R_0, Sb(A_2) \cup X\big) = \overrightarrow{\mathfrak{M}_2}(X)$, *for every* $X \subseteq S_2$.

By Theorems 2.38 and 3.22 (ii) we get one of the most important theorems on classical propositional logic, namely the substitutional version of classical propositional logic is Post-complete, see [65], 1929.

Theorem 3.39. $\langle R_{0*}, A_2 \rangle \in \mathrm{Cpl}$.

On the other hand, it is easy to see that the invariant system of this logic is Post-incomplete:

Theorem 3.40. $dc\big(R_0, Sb(A_2)\big) = \mathfrak{c}$.

Proof. By 3.33, we know that every $Y \in \mathscr{L}\big(R_0, Sb(A_2)\big)$ determines a valuation $v \colon At \to \{0,1\}$ such that $Y = h^{v^{-1}}(\{1\})$. On the other hand, every set of the form $h^{v^{-1}}(\{1\})$ is a Lindenbaum overset of $Cn\big(R_0, Sb(A_2)\big)$. Two different valuations determine different Lindenbaum oversets, thus $dc\big(R_0, Sb(A_2)\big) = \mathfrak{c}$. $\qquad\square$

Of course, the degree of completeness of any logic weaker than $\langle R_0, Sb(A_2) \rangle$ also is equal to $\mathfrak{c}$. Moreover, (see [4], 1981),

Corollary 3.41. $\mathscr{L}\big(R_0, \bigcap\{Y_i : i \leqslant n\}\big) = \{Y_1, \ldots, Y_n\}$ *if* $\{Y_i\}_i \subseteq \mathscr{L}\big(R_0, Sb(A_2)\big)$.

Proof. Inclusion $(\supseteq)$ is obvious. Assume $Y \in \mathscr{L}\big(R_0, \bigcap\{Y_i : i \leqslant n\}\big) \setminus \{Y_1, \ldots, Y_n\}$. Hence $Y \in \mathscr{L}\big(R_0, Sb(A_2)\big)$ and there are formulas $\alpha_i \in Y \setminus Y_i$ $(i \leqslant n)$. So, $Cn(R_0, Y_k \cup \{\alpha_i\}_{i \geqslant 1}) = S_2$ for each $k \leqslant n$. Then $\sim (\alpha_1 \cdot \ldots \cdot \alpha_n) \in Cn(R_0, Y_k) = Y_k$ and hence $\sim (\alpha_1 \cdot \ldots \cdot \alpha_n) \in Cn(R_0, \bigcap\{Y_i : i \leqslant n\}) \subseteq Y$ which is impossible. $\qquad\square$

3.3 The problem of uniqueness of Lindenbaum extensions

In the previous section we considered Lindenbaum supersets and oversystems of a given propositional logic. These constructions are vaguely called Lindenbaum extensions. An interesting problem is to count the number of such extensions, that is to find the cardinality of the family $\mathscr{L}(R,X)$ for a given system $\langle R,X\rangle$. Particularly, one can look for such logics $\langle R,X\rangle$ for which $Nc(\mathscr{L}(R,X)) = 1$. Let us assume the following definition (see [4], 1976):

Definition 3.42. $\langle R,X\rangle \in \mathscr{T}(M) \Leftrightarrow \mathscr{L}(R,X) = \{M\}$, for $M \subseteq S$.

We shall say that $\langle R,X\rangle$ has *Tarski's property* iff $\langle R,X\rangle \in \mathscr{T}(M)$ for some $M \neq S$. In this section we shall deal with results concerning the $\mathscr{T}(M)$-property.

Intermediate logics

Let us consider the intuitionistic propositional logic defined in Chapter 1. We recall that the consequence operation generated by the invariant version of this logic $\langle R_0, Sb(A_i)\rangle$ is denoted by Cn_i. We will also consider some extensions and fragments of the intuitionistic logic; in the first place the positive fragment, which is the so-called Hilbert's logic $\langle R_0, Sb(A_H)\rangle$ where $A_H = S_1 \cap A_i$. One feature of the intuitionistic (likewise Hilbert's) logic is the validity of the deduction theorem and, consequently, consistency theorem (see Corollary 1.66);

Lemma 3.43. *For every $X \subseteq S_2$ and every $\alpha, \beta \in S_2$,*

(i) $(\alpha \to \beta) \in Cn_i(X) \Leftrightarrow Cn_i(X \cup \{\beta\}) \subseteq Cn_i(X \cup \{\alpha\})$;

(ii) $Cn_i(X \cup \{\alpha\}) \neq S_2 \quad \Leftrightarrow \quad \sim\alpha \notin Cn_i(X)$.

The intuitionistic logic is weaker than the classical one. Nevertheless, the same characterization of consistent and complete extensions are valid for the intuitionistic logic as in the classical case, see Corollary 3.29;

Corollary 3.44. *For each $X \subseteq S_2$,*

(i) $\langle R_0, Sb(A_i) \cup X\rangle \in \text{Cns} \Leftrightarrow (\neg(\alpha, \sim\alpha \in Cn_i(X))$, *for some* $\alpha \in S_2)$;

(ii) $\langle R_0, Sb(A_i) \cup X\rangle \in \text{Cpl} \Leftrightarrow (\alpha \in Cn_i(X) \vee \sim\alpha \in Cn_i(X)$, *for each* $\alpha \in S_2)$.

Corollary 3.45.

(i) $\mathscr{L}(R_0, Sb(A_2)) = \mathscr{L}(R_0, Sb(A_i))$;

(ii) $\mathscr{L}(R_{0*}, A_2) = \mathscr{L}(R_{0*}, A_i) = \{Z_2)\}$.

Proof. (i): Inclusion ($\subseteq$) is obvious; to prove the reverse inclusion ($\supseteq$) it suffices to show that $Sb(A_2) \subseteq Y$ for every $Y \in \mathscr{L}(R_0, Sb(A_i))$. By Glivenko's theorem, Corollary 2.91, we have $\sim\sim\alpha \in Cn(R_0, Sb(A_i)) \subseteq Y$ for each $\alpha \in Sb(A_2)$. Hence, we get $\sim\alpha \notin Y$ and consequently, by 3.44, $\alpha \in Y$.

(ii): Inclusion ($\subseteq$) is obvious; assume $Y \in \mathscr{L}(R_{0*}, A_i) \setminus \mathscr{L}(R_{0*}, A_2)$. We have $Cn(R_{0*}, A_2 \cup Y) = S_2$. By Lindenbaum's theorem there is $Y_1 \in \mathscr{L}(R_0, Y \cup Sb(A_i))$ and hence using 3.45 we get $Sb(A_2) \subseteq Y_1$. Thus, we reach a contradiction as $S_2 = Cn(R_{0*}, A_2 \cup Y) \subseteq Cn(R_0, A_i \cup Y_1) \neq S_2$.   □

This corollary implies that for the system $\langle R_0, Sb(A_i) \rangle$ there does not hold a counterpart of Lindenbaum–Łoś's theorem: let $\beta \in S_2$ be a formula such that $\beta \in Cn(R_{0*}, A_2) \setminus Cn(R_{0*}, A_i)$, then $\mathscr{L}^\beta(R_0, Sb(A_i)) \cap \mathscr{L}(R_0, Sb(A_2)) = \emptyset$. Thus, there exist relative Lindenbaum (R_0-closed) oversets for the set $Cn(R_0, Sb(A_i))$ which are not Lindenbaum supersets. A relative Lindenbaum superset is a Lindenbaum superset iff it contains all axioms of the classical logic (this follows directly from Corollaries 3.30 and 3.45). The above concerns, of course, R_0-closed sets. We can also formulate this fact in the following form (see [6], 1974):

Lemma 3.46. *If $Y \in \mathscr{L}^\beta(R_0, Sb(A_i))$, then*

$$\langle R_0, Y \rangle \notin \mathrm{Cpl} \quad \Leftrightarrow \quad ((\beta \to \alpha) \to \beta \in Y, \quad \text{for some } \alpha \in S_2).$$

Proof. Let us assume that $\langle R_0, Y \rangle \notin \mathrm{Cpl}$. Then, there is a formula $\gamma \in S$ such that $\beta \in Cn(R_0, Y \cup \{\gamma\}) \neq S_2$. This means that $Cn(R_0, Y \cup \{\beta\}) \neq S_2$ and hence $\beta \to \alpha \notin Y$ for some $\alpha \in S_2$. Thus, we have $(\beta \to \alpha) \to \beta \in Y$.

Now, assume that $(\beta \to \alpha) \to \beta \in Y$ for some $\alpha \in S_2$ and let $\langle R_0, Y \rangle \in \mathrm{Cpl}$. Then $Cn(R_0, Y \cup \{\beta\}) = S_2$, and hence by the deduction theorem $\beta \to \alpha \in Y$ which is impossible.   □

Most theorems proved in this section for propositional systems in the language S_2, hold also for any implicative sublanguage. In particular, by the same argument as in the proof of Lemma 3.46, we get the above equivalence for every $Y \in \mathscr{L}^\beta(R_0, Sb(A_H))$. Thus, we get for Hilbert's logic,

Corollary 3.47. $\mathscr{L}(R_0, Sb(A_H)) = \mathscr{L}(R_0, Sb(A_2^{\to + \cdot}))$.

Proof. The inclusion ($\supseteq$) is obvious. Now, let us assume that $Y \in \mathscr{L}(R_0, Sb(A_H))$. We will consider the following two possibilities:
   1. $\beta \notin Y$. Then we get $((\beta \to \alpha) \to \beta) \to \beta \in Y$ for each formula $\alpha \in S_1$ $\left(\text{since } Y \in \mathscr{L}^\beta(R_0, Sb(A_H)) \right)$.
   2. $\beta \in Y$. Then from the fact that $\beta \to (((\beta \to \alpha) \to \beta) \to \beta) \in Y$ if follows that $((\beta \to \alpha) \to \beta) \to \beta \in Y$. Therefore $Sb(\{((p \to q) \to p) \to p\}) \subseteq Y$ which means that $Sb(A_2^{\to + \cdot}) \subseteq Y$.   □

By Theorem 3.40 and Corollaries 3.45 and 3.47 we can determine the degree of completeness of any consistent system $\langle R_0, Sb(A_H) \cup X \rangle$. Since such logics possess relative Lindenbaum oversystems which are not Lindenbaum oversystems, then it seems interesting to establish the cardinality of the family of all relative Lindenbaum oversystems for any formula $\beta \notin Cn(R_0, Sb(A_H) \cup X)$. The following theorem holds for any implicative sublanguage of S_2 (see [6], 1974):

Theorem 3.48. *If $p_0 \to p_0 \in X$ and $\alpha \notin Cn(R_0, Sb(X))$, then*

$$Nc\big(\mathscr{L}^\alpha(R_0, Sb(X))\big) = \mathfrak{c}.$$

Proof. At first, we prove that $\alpha \notin Cn(R_0, Sb(X) \cup A \cup \{p_i \to \alpha : p_i \notin A \cup At(\alpha)\})$ for every $A \subseteq At \setminus At(\alpha)$. Suppose, on the contrary, it is not true. Then for the substitution $e \colon At \to S_2$ defined by the clause

$$e(p_i) = \begin{cases} p_i \to p_i, & \text{if } p_i \in A \\ \alpha, & \text{if } p_i \notin A \cup At(\alpha) \\ p_i, & \text{if } p_i \in At(\alpha) \end{cases}$$

we get $\alpha = h^e(\alpha) \in Cn\big(R_0, Sb(X) \cup h^e(A) \cup h^e\{p_i \to \alpha : p_i \notin A \cup At(\alpha)\}\big) = Cn(R_0, Sb(X))$ which is impossible.

From Theorem 3.26 we infer that there is a set Y such that

$$Y \in \mathscr{L}^\alpha\big(R_0, Sb(X) \cup A \cup \{p_i \to \alpha : p_i \notin A \cup At(\alpha)\}\big).$$

Moreover, we have $Y \in \mathscr{L}^\alpha(R_0, Sb(X))$ and $At \cap Y \setminus At(\alpha) = A$. Thus, one may define, taking $f(Y) = At \cap Y \setminus At(\alpha)$, a mapping f from $\mathscr{L}^\alpha(r_0, Sb(X))$ onto the set $2^{At \setminus At(\alpha)}$. Since the set $At \setminus At(\alpha)$ is infinite, we conclude that $Nc\big(\mathscr{L}^\alpha(R_0, Sb(X))\big) = \mathfrak{c}$. $\qquad\square$

The above theorem comprises also, as its particular case, the invariant system of the classical propositional logic. By Theorem 3.30, we again come to the conclusion that the degree of completeness of the system $\langle R_0, Sb(A_2) \rangle$ is $\mathfrak{c}$.

It is easy to see now that the question concerning the degree of completeness is essential only if one considers propositional systems with substitution as a derivable rule. It is so because the degree of completeness for the majority of invariant logics is equal to $\mathfrak{c}$.

Now we determine the degree of completeness of the intuitionistic logic and its certain extensions (see [149], 1974; [16], 1959):

Theorem 3.49.

(i) $dc(\langle R_{0*}, A_i \rangle) = dc(\langle R_{0*}, A_H \rangle) = \mathfrak{c}$;

(ii) $dc(\langle R_{0*}, A_l \rangle) = \aleph_0$;

(iii) $dc(\langle R_{0*}, H_n \rangle) = n$, *for each* $n \geqslant 2$.

The above easily follows from the appropriate completeness theorems for the logics considered.

Many-valued logics of Łukasiewicz

Łukasiewicz's many-valued logics have many non-standard properties. However, there are also known for them variants of the deduction theorem and variants of Tarski's lemma on consistency, see Theorems 1.62 and 1.65. Hence we can get for them a similar body of results, as concerns Lindenbaum extensions, as we got for the intuitionistic logic. In particular, we have

Corollary 3.50. $\mathscr{L}(R_{0*}, Ł_\infty) = \mathscr{L}(R_{0*}, Ł_n) = \mathscr{L}(R_{0*}, A_2) = \{Z_2\}$.

Moreover, we can easily put bounds on the degree of completeness of the considered logics (see [110], 1952):

Theorem 3.51.

  (i) $dc(\langle R_{0*}, Ł_\infty \rangle) \geqslant \aleph_0$;

  (ii) $dc(\langle R_{0*}, Ł_n \rangle) \leqslant 2^{k-1} + 1$, *where k is the number of divisors of $n - 1$.*

Now, we pass to the main subject of this chapter. It is easy to see that Lindenbaum's theorem (Lindenbaum–Asser's theorem) does not determine, in general, a unique complete superset (oversystem) for a given set of theorems $Cn(R, X)$ (of a given system $\langle R, X \rangle$).

Example. Let us consider the system $\langle R_{0*}, \{p \to p\} \rangle$. A Lindenbaum oversystem for this system is, of course, the classical logic $\langle R_{0*}, A_2^{\to} \rangle$. Now, for the system $\langle R_{0*}, \{p \to p, (p \to q) \to (q \to p)\} \rangle$ there exists, according to Lindenbaum's theorem, a Lindenbaum superset $Y \supseteq Cn(R_{0*}, \{p \to p, (p \to q) \to (q \to p)\})$. The set Y is at the same time the Lindenbaum overset for $Cn(R_{0*}, \{p \to p\})$. But, of course, $Y \neq Cn(R_{0*}, A_2^{\to})$. Then for the system $\langle R_{0*}, \{p \to p\} \rangle$ there exist at least two different Lindenbaum oversystems $\langle R_{0*}, A_2^{\to} \rangle$ and $\langle R_{0*}, Y \rangle$.

On the other hand, there exist propositional logics which have exactly one L-extension. For instance, the only L-extension for the logics $\langle R_{0*}, A_i \rangle$ and $\langle R_{0*}, Ł_\infty \rangle$ is the classical logic $\langle R_{0*}, A_2 \rangle$. Both $\langle R_{0*}, A_i \rangle$ and $\langle R_{0*}, Ł_\infty \rangle$ possess then Tarski's property (see Definition 3.42). More formally, we have

$$\langle R_{0*}, A_i \rangle \in \mathscr{T}(Z_2) \quad \text{and} \quad \langle R_{0*}, Ł_\infty \rangle \in \mathscr{T}(Z_2),$$

where $Z_2 = Cn(R_{0*}, A_2)$. In this situation one can ask how strong a propositional system ought to be (we confine ourselves to R_{0*}-systems) so as to have as its only L-extension the system $\langle R_{0*}, A_2 \rangle$. This question is interesting and important not only from the purely formal point of view. Namely, we have proved that the system $\langle R_{0*}, \{p \to p\} \rangle$ has at least two different L-extensions $\langle R_{0*}, A_2 \rangle$ and $\langle R_{0*}, Y \rangle$ (see the above example). In the first system the symbol $\to$ occurring in the initial formula ('axiom') $p \to p$ remains to be the implication connective, whereas in the second system $\langle R_{0*}, Y \rangle$ the symbol $\to$ cannot be interpreted as the implication connective. So we can rightly think that systems with only one

L-extension determine their primitive connectives in a unique manner or, in other words, determine their connectives properly.

Let $A_T = \{p \to (q \to p), \, p \to \sim\sim p, \sim p \to (p \to q), \, p \to (\sim q \to \sim (p \to q)),$
$p \to (p+q), q \to (p+q), \sim p \to (\sim q \to \sim (p+q)), \, p \to (q \to p \cdot q), \sim p \to \sim (p \cdot q),$
$\sim q \to \sim (p \cdot q)\}$. Tarski (see [119], 1935) has proved that $\langle R_{0*}, A_2 \rangle$ is the only complete and consistent extension for $\langle R_{0*}, A_T \rangle$. In symbols, $\mathscr{L}(R_{0*}, A_T) = \{Z_2\}$. Some further results concerning the cardinality of families of L–extensions were obtained a few years after Tarski's result. But this problem was neither exactly examined, nor Tarski's theorem essentially generalized before Biela [4],1976.

Let us recall that $\mathscr{T}(M)$ (see Definition 3.42) is the family of systems $\langle R, X \rangle$ such that $\langle R, M \rangle$ is the only L-extension for $\langle R, X \rangle$.

Lemma 3.52. *If $\langle R, A \rangle \in \text{Comp} \cap \text{Cns}$ and $M \neq S$, then $\langle R, A \rangle \in \mathscr{T}(M)$ iff*

$$Cn(R, A \cup M) = M \, \wedge \, Cn(R, A \cup \{\alpha\}) = S, \quad \text{for each } \alpha \notin M.$$

Easy proof of this lemma is left to the reader.

Let us define a substitution: $\bar{e}(p_i) = (p \to p) \to \sim\sim (p \to p)$ for each $i \in \mathbb{N}$. Then we get (see [119], 1935)

Lemma 3.53. *For each formula $\alpha \in S_2$ with $At(\alpha) = \{p\}$:*

(i) $h^{\bar{e}}(\alpha) \in Z_2 \quad \Rightarrow \quad h^{\bar{e}}(\alpha) \in Cn(R_{0*}, A_T)$;

(ii) $h^{\bar{e}}(\alpha) \notin Z_2 \quad \Rightarrow \quad \sim h^{\bar{e}}(\alpha) \in Cn(R_{0*}, A_T)$.

Easy inductive proof can be omitted. Let us prove

Theorem 3.54.

(i) *There are propositional systems $\langle R_{0*}, X \rangle \in \mathscr{T}(Z_2)$ weaker than the system $\langle R_{0*}, A_T \rangle$ as given by Tarski;*

(ii) *For every $\langle R_{0*}, X \rangle \in \mathscr{T}(Z_2)$ the set $Cn(R_{0*}, X) \cap Cn(R_{0*}, A_T)$ is non-empty.*

Proof. (i): Let $p \in At$. Define $S_p = \{\alpha \in S_2 : At(\alpha) = \{p\}\}$ and let us take $A = S_p \cap Cn(R_{0*}, A_T)$. We will show that $\langle R_{0*}, A \rangle \in \mathscr{T}(Z_2)$. By the completeness of the classical logic with respect to the two-valued matrix $\mathfrak{M}_2$, for each $\alpha \notin Z_2$ one can find a mapping $v \colon At \to \{0,1\}$ such that $h^v(\alpha) = 0$. Let us define a substitution $e \colon At \to S_2$ as

$$e(p_i) = \begin{cases} p \to p, & \text{if } v(p) = 1 \\ \sim (p \to p), & \text{if } v(p) = 0. \end{cases}$$

Of course, we have $At(h^e(\alpha)) = \{p\}$ and $\sim h^e(\alpha) \in Z_2$. Thus, by Lemma 3.53, we get $\sim h^{\bar{e}} h^e(\alpha) \in A$. Since $\sim p \to (p \to q) \in A_T$, then $(h^{\bar{e}} h^e(\alpha) \to p) \in A$. So, we have $S_2 = Cn(R_{0*}, A \cup \{\alpha\})$. To prove that $Cn(R_{0*}, A) \neq Cn(R_{0*}, A_T)$ it suffices to observe first that $Cn(R_{0*}, A) \subseteq Cn(R_{0*}, Z_2 \cap S_p) \subseteq Sb(Z_2 \cap S_p)$ and then $p \to (q \to p) \in Cn(R_{0*}, A_T) \setminus Sb(Z_2 \cap S_p)$.

(ii): Let $\alpha \in Cn(R_{0*}, X) \cap S_p$, where $\langle R_{0*}, X \rangle \in \mathscr{T}(Z_2)$. Then, we have $h^{\bar{e}}(\alpha) \in Cn(R_{0*}, X) \subseteq Z_2$ and by Lemma 3.53, $h^{\bar{e}} \in Cn(R_{0*}, A_T)$. $\square$

Clearly, each consistent $\langle R_{0*}, X\rangle$ stronger than any $\langle R_{0*}, A\rangle \in \mathscr{T}(M)$ is also an element of $\mathscr{T}(M)$. Hence, it follows from the proof of the above theorem that $\langle R_{0*}, A_T\rangle \in \mathscr{T}(Z_2)$. Moreover, let us note that 3.54 (ii) does not extend on systems $\langle R, A\rangle$ with $R \neq R_{0*}$. To show this take $r = \{\langle \alpha, \beta\rangle \in S^2 : \alpha \notin S_2 \setminus Z_2 \wedge \beta \in S_2\}$ and notice that $\langle \{r\}, \emptyset\rangle \in \mathscr{T}(Z_2)$.

Vaguely speaking, Theorem 3.53(ii) shows that there are systems $\langle R_{0*}, X\rangle$ with Tarski's property different from $\langle R_{0*}, A_T\rangle$ but sets of the theorems of these systems must have non-empty intersections with $Cn(R_{0*}, A_T) = Z_T$. Thus, looking for various families of systems with $\mathscr{T}(Z_2)$-property, we can encounter six situations in which the elements of such a family are related to the system $\langle R_{0*}, A_T\rangle$ (and to themselves). These situations can be formally defined but we represent them with the following drawings:

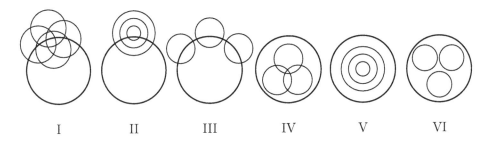

$$\text{I} \qquad\qquad \text{II} \qquad\qquad \text{III} \qquad\qquad \text{IV} \qquad\qquad \text{V} \qquad\qquad \text{VI}$$

For each drawing there was found, in [4], a family with maximal cardinality. Here, we present only a family represented by the drawing I.

Let us define the relation $\#$ between R-systems: $\langle R, X\rangle \# \langle R, X_1\rangle$, in words: the systems *overlap* , iff the sets $Cn(R, X) \cap Cn(R_1, X_1)$, $Cn(R, X) \setminus Cn(R_1, X_1)$ and $Cn(R, X_1) \setminus Cn(R, X)$, are non-empty. Then the figure I represents a family such that every system from this family overlaps with $\langle R_{0*}, A_T\rangle$ and, if $\langle R, X\rangle$, $\langle R, X_1\rangle$ both belong to this family, then they also overlap.

Let S_{01} be the set of all formulas built by means of $p \to p$ and $\sim (p \to p)$ as atoms. Formally, we have $S_{01} = h^e(S_2)$ for any $e \colon At \xrightarrow{onto} \{p \to p, \sim (p \to p)\}$. It is known that for every $\alpha \in S_{01}$ we have $\alpha \in Z_2$ or $\sim \alpha \in Z_2$. Define now

$$\Delta_p = \{\alpha \to p : \alpha \in S_{01}\} \cap Z_2,$$
$$\Delta_{\sim(p\to p)} = \{\alpha \to\sim (p \to p) : \alpha \in S_{01}\} \cap Z_2.$$

Lemma 3.55.

(i) $\langle R_{0*}, \Delta_p\rangle \in \mathscr{T}(Z_2)$;

(ii) $\langle R_{0*}, \Delta_p\rangle \# \langle R_{0*}, A_T\rangle \# \langle R_{0*}, \Delta_p \cup \Delta_{\sim(p\to p)}\rangle$;

(iii) *If* $X \subseteq \Delta_{\sim(p\to p)}$, *then* $Cn(R_{0*}, \Delta_p \cup X) = Sb(\Delta_p) \cup Sb(X)$.

Proof. (i): For every $\alpha \notin Z_2$ there exists $e \colon At \to \{p \to p, \sim (p \to p)\}$ such that $h^e(\alpha) \notin Z_2$. Then $h^e(\alpha) \to p \in \Delta_p$ and consequently $Cn(R_{0*}, \Delta_p \cup \{\alpha\}) = S_2$ which by Lemma 3.52 ends the proof.

(ii): We have $\sim (p \to p) \to p \in \Delta_p \setminus Cn(R_{0*}, A_T)$ which can be proved by use of the matrix $\mathfrak{M} = \langle \{0, 1, 2\}, \{1\}, f^\to, f^\sim, f^+, f^\cdot \rangle$ where

$f^\to$	0	1	2		$f^\sim$			$f^\cdot$	0	1	2		f^+	0	1	2
0	2	1	1		0	2		0	0	1	2		0	1	1	2
1	2	1	2		1	2		1	1	1	2		1	1	1	1
2	1	1	1		2	1		2	2	2	2		2	2	1	2

Moreover, $Cn(R_{0*}, \Delta_p \cup \Delta_{\sim(p \to p)}) \subseteq Sb(Z_2 \cap S_p) \not\ni p \to (q \to p)$ and, by Theorem 3.54 (ii), $Cn(R_{0*}, \Delta_p) \cap Cn(R_{0*}, A_T) \neq \emptyset$.

(iii): One ought to prove that $r_0 \in \mathrm{Adm}(r_*, X \cup \Delta_p)$; if $\alpha \to \beta \in Sb(X \cup \Delta_p)$, then of course $\sim \alpha \in Z_2$, thereby $\alpha \notin Sb(X \cup \Delta_p)$. In other words: r_0 is admissible since it does not work on this set. □

Theorem 3.56. *There is an uncountable family $\mathscr{R}$ of systems $\langle R_{0*}, X \rangle$ (where $X \subseteq S_2$) such that:*

(i) *if $\langle R_{0*}, X \rangle \in \mathscr{R}$, then $\langle R_{0*}, X \rangle \in \mathscr{T}(Z_2)$;*

(ii) *if $\langle R_{0*}, X_1 \rangle, \langle R_{0*}, X_2 \rangle \in \mathscr{R}$ and $X_1 \neq X_2$, then $\langle R_{0*}, X_1 \rangle \# \langle R_{0*}, X_2 \rangle$;*

(iii) *if $\langle R_{0*}, X \rangle \in \mathscr{R}$, then $\langle R_{0*}, X \rangle \# \langle R_{0*}, A_T \rangle$.*

Proof. Let $\delta_n = (p \to p) \xrightarrow{n} \sim (p \to p)$ $(n \geqslant 1)$. Then $\delta_n \to \sim (p \to p) \in \Delta_{\sim(p \to p)}$ and $Sb(\delta_n \to \sim (p \to p)) \cap Sb(\Delta_p) = \emptyset$ for each $n \in \mathbb{N}$). Moreover, one easily checks that $Sb(\{\delta_n \to \sim (p \to p)\}) \cap Sb(\{\delta_m \to \sim (p \to p)\}) = \emptyset$ if $n \neq m$. Thus, by Lemma 3.55, we conclude that $\bigcup_{n \geqslant 1} \{\delta_n \to \sim (p \to p)\}$ is an independent set with respect to the consequence operation generated by $\langle R_{0*}, \Delta_p \rangle$. Since for every infinite countable set there exists a family $\mathscr{R}$ of its subsets such that $Nc(\mathscr{R}) = \mathfrak{c}$ and $X_1 \setminus X_2 \neq \emptyset$, $X_1 \cap X_2 \neq \emptyset$, $X_2 \setminus X_1 \neq \emptyset$ for each $X_1, X_2 \in \mathscr{R}$, we conclude that $\{\langle R_{0*}, \Delta_p \cup X \rangle : X \subseteq \bigcup_{n \geqslant 1} \{\delta_n \to \sim (p \to p)\}\}$ contains the desired family. □

Let us note, moreover, that all systems considered in the proof of 3.56 are not finitely axiomatizable. It can be easily verified that for each $X \subseteq \Delta_p$ and each $Y \subseteq \Delta_{\sim(p \to p)}$ we have $Cn(R_{0*}, X \cup Y) = Sb(X) \cup Sb(Y)$ and, for non-finite X and Y the inclusion $\Delta_p \subseteq Sb(X) \cup Sb(Y)$ holds.

The above theorem solves the question concerning the number of R_{0*}-systems with Tarski's property. Note that families represented by the figures III and IV can be infinite but countable. From the mere existence of two separate systems with $\mathscr{T}(Z_2)$-property it follows, of course, that a least R_{0*}-system with the $\mathscr{T}(Z_2)$-property does not exist.

3.4   Structural completeness

In this chapter we again investigate the notion of generalized completeness, Γ-Cpl (see Definition 3.1). Theorem 3.15 states that every admissible rule of a Post-complete logic is also a derivable rule of this logic. By Theorem 1.60,

Lemma 3.57. *If* $\langle R, A \rangle \in \mathrm{Inv}$ *and* $A \neq \emptyset$, *then* $r_* \in \mathrm{Der}(R, A) \Leftrightarrow \langle R, A \rangle \notin \mathrm{Cns}$.

The substitution rule is then admissible and non-derivable in almost all invariant consistent logics (the exceptions are systems $\langle R, \emptyset \rangle$ such that $Cn(R, \{\alpha\}) = S$ for each $\alpha \in S$). By 3.57 we get, of course, $\langle R_0, Sb(A_2) \rangle \notin \mathrm{Cpl}$. Nevertheless, it can happen that the set of derivable rules of a given Post-incomplete system contains all admissible (in this system) and structural rules. The set of structural rules constitutes an important subset of all inferential rules, therefore the situation presented above seems to be essential enough to introduce a new notion describing it. This notion is named *structural completeness* (and was introduced in [79], 1971).

If $\Gamma = S$, we shall write – see Definition 3.1(i) – SCpl instead of S-Cpl:

$$\langle R, A \rangle \in \mathrm{SCpl} \quad \Leftrightarrow \quad \mathrm{Adm}(R, A) \cap \mathrm{Struct} \subseteq \mathrm{Der}(R, A),$$

for $A \subseteq S$, $R \subseteq \mathscr{R}_S$. Of course, we have $\mathrm{Cpl} \neq \mathrm{SCpl}$ and, by Lemma 3.10, we get the following characterization of structural completeness:

Corollary 3.58.

$$\langle R, A \rangle \in \mathrm{SCpl} \Leftrightarrow \forall_{R_1 \subseteq \mathrm{Struct}} \forall_{A_1 \subseteq S} \, [Cn(R, A) = C(R_1, A_1) \Rightarrow \langle R_1, A_1 \rangle \preccurlyeq \langle R, A \rangle].$$

If we restrict ourselves to invariant logics, then we get (see [68], 1976)

Corollary 3.59. *If* $\langle R, A \rangle \in \mathrm{Inv}$, *then*

$$\langle R, A \rangle \in \mathrm{SCpl} \Leftrightarrow \forall_{\langle R_1, A_1 \rangle \in \mathrm{Inv}} \, [Cn(R, A) = Cn(R_1, A_1) \Rightarrow \langle R_1, A_1 \rangle \preccurlyeq \langle R, A \rangle].$$

An invariant propositional system is then structurally complete iff it is maximal in the family of all invariant systems with the fixed set of theorems. We can write down 3.59 in the consequence formalism:

Corollary 3.59 (∞). *If* $Cn \in \mathrm{STRUCT}$, *then*

$$Cn \in \mathrm{SCPL} \Leftrightarrow \forall_{Cn_1 \in \mathrm{STRUCT}} \, [Cn_1(\emptyset) = Cn(\emptyset) \Rightarrow Cn_1 \leqslant Cn.$$

By Theorem 3.4, for $\Gamma = S$, we get the following two corollaries (see [84] 1973).

Corollary 3.60. *If* $r_* \in \mathrm{Adm}(R, A)$ *and* $\emptyset \neq Cn(R, A) \neq S$, *then*

$$\mathrm{Der}(R, A) \cap \mathrm{Struct} = V(\mathfrak{M}) \cap \mathrm{Struct} \Leftrightarrow Cn(R, A) = E(\mathfrak{M}) \wedge \langle R, A \rangle \in \mathrm{SCpl}.$$

Corollary 3.61. *If $Cn(R, A) = E(\mathfrak{M})$, then*

$$\langle R, A \rangle \in \text{SCpl} \quad \Leftrightarrow \quad V(\mathfrak{M}) \cap \text{Struct} \subseteq \text{Der}(R, A).$$

Corollary 3.61 states an intuitively important fact that if $\mathfrak{M}$ is an adequate matrix for $\langle R, A \rangle$, then $\langle R, A \rangle$ is structurally complete iff all structural and $\mathfrak{M}$-valid rules are derivable in $\langle R, A \rangle$.

Of course, not all systems are structurally complete but by Theorem 3.13 we get

Corollary 3.62. *Every consistent system $\langle R, A \rangle$ can be consistently extended to a structurally complete system $\langle R_1, A_1 \rangle$.*

Structural completeness for systems over S_2 is — rather unexpectedly — connected with Tarski's property (see [5], 1982):

Theorem 3.63. *If $\langle R, A \rangle \in \text{Inv}$ and $Cn(R, A \cup Z_2) \subseteq Z_2$, then*

$$\langle R \cup \{r_*\}, A \rangle \in \text{SCpl} \quad \Rightarrow \quad \langle R \cup \{r_*\}, A \rangle \in \mathcal{T}(Z_2).$$

Proof. Let us assume that $\alpha \notin Z_2$. By the completeness theorem, we have $h^v(\alpha) = 0$ for some $v \colon At \to \{0, 1\}$. Define $e \colon At \to S_2$ as follows:

$$e(\gamma) = \begin{cases} p \to p & \text{if } v(\gamma) = 1 \\ \sim (p \to p) & \text{if } v(\gamma) = 0. \end{cases}$$

Of course, $\sim h^e(\alpha) \in Z_2$. Consider the rule $r = \{\langle \phi, \psi \rangle : \phi \in Sb(h^e(\alpha))\}$. Obviously, $r \in \text{Struct} \cap \text{Adm}(R, A)$. Thus, $r \in \text{Der}(R \cup \{r_*\}, A)$. For every $\psi \in S_2$ we have $\psi \in Cn(\{r\}, \{h^e(\alpha)\}) \subseteq Cn(R \cup \{r_*\}, A \cup \{\alpha\}) = S_2$. We have proved that, for every $Y \subseteq S$ such that $Cn(R \cup \{r_*\}, A \cup Y) \neq S_2$, we have $Y \subseteq Z_2$ and hence $\mathscr{L}(R \cup \{r_*\}, A) = \{Z_2\}$. $\square$

If $R \subseteq \text{Struct}$ and $Cn(R, A \cup Z_2) \subseteq Z_2$, then $R \subseteq \text{Der}(R_{0*}, A_2)$. Thus, the intuitive meaning of 3.63 is that a structurally complete substitutional subsystem of $\langle R_{0*}, A_2 \rangle$ determines uniquely its logical connectives. In view of Corollary 3.60 and Theorem 3.63 one can say that structurally complete logic is a good logic.

Using the notion of the 'big rule' (see Lemma 3.8) we can get a simple characterization of structural completeness ([98], 1972):

Corollary 3.64. $\langle R, A \rangle \in \text{SCpl} \quad \Leftrightarrow \quad r_{Cn(R,A)} \in \text{Der}(R, A).$

We can also strengthen Corollary 3.6 to the following result:

Lemma 3.65. $\langle R, A \rangle \in \text{SCpl} \Leftrightarrow N(\mathfrak{M}^{R,A}) \subseteq \text{Der}(R, A).$

Proof. The proof is by Lemma 2.44 and Corollary 3.64. $\square$

It is also possible to write down this lemma in the matrix-consequence formalism

Corollary 3.66 (∞). $\langle R, A \rangle \in \mathrm{SCpl} \Leftrightarrow \overrightarrow{(\mathfrak{M}^{R,A}}(X) \subseteq Cn_{RA}(X)$, *for each* $X \subseteq S$).

For invariant logics the above characterization of structural completeness becomes more visible as we get:

Corollary 3.67. *If* $\langle R, A \rangle \in \mathrm{Inv}$, *then*

      (i) $\langle R, A \rangle \in \mathrm{SCpl} \Leftrightarrow N(\mathfrak{M}^{R,A}) = \mathrm{Der}(R, A)$;

      (ii)   (∞) $\langle R, A \rangle \in \mathrm{SCpl} \Leftrightarrow \forall_{X \subseteq S} \; \overrightarrow{\mathfrak{M}^{R,A}}(X) = Cn(R, A \cup X)$;

      (iii) (fin) $\langle R, A \rangle \in \mathrm{SCpl} \Leftrightarrow \forall_{X \in \mathrm{Fin}(S)} \; \overrightarrow{\mathfrak{M}^{R,A}}(X) = Cn(R, A \cup X)$.

There exists an interesting characterization of structural completeness for invariant systems which is quite different from the preceding ones (see [105], 1977).

Theorem 3.68 (∞). *Let* $\langle R, A \rangle$ *be an invariant logic such that* Cn_{RA} *is finitistic. Then* $\langle R, A \rangle \in \mathrm{SCpl}$ *iff for every* $\psi \in S$ *and for every* $Y \in \mathscr{L}^\psi(R, A)$ *there is an endomorphism* $h \colon S \to S$ *such that* $Y = h^{-1}\big(Cn(R, A)\big)$.

Proof. ($\Rightarrow$): Suppose this implication is not true, i.e., $\langle R, A \rangle \in \mathrm{SCpl}$ and

($\star$) $h^e(Y) \subseteq Cn(R, A) \quad \Rightarrow \quad Y \neq h^{e^{-1}}\big(Cn(R, A)\big)$,

for all $e \colon At \to S$ and for some $Y \in \mathscr{L}^\psi(R, A)$. Suppose that $h^e(Y) \subseteq Cn(R, A)$ for a substitution $e \colon At \to S$. By ($\star$), we get $\psi \in Cn(R, A \cup Y \cup \{\phi\})$ for some $\phi \in h^{e^{-1}}\big(Cn(R, A)\big) \setminus Y$. Then $h^e(\psi) \in Cn\big(R, A \cup h^e(Y) \cup \{h^e(\phi)\}\big) \subseteq Cn(R, A)$. Thus, we have proved that for each $e \colon At \to S$,

$$h^e(Y) \subseteq Cn(R, A) \quad \Rightarrow \quad h^e(\psi) \in Cn(R, A).$$

Hence, by Corollary 3.64, we get $\psi \in Cn(R, Y)$ which is impossible.

    ($\Leftarrow$): Assume that $r \in \mathrm{Struct} \setminus \mathrm{Der}(R, A)$. Then, there exists $\langle \Pi, \psi \rangle \in r$ such that $\psi \notin Cn(R, A \cup \Pi)$. Suppose that $Y \in \mathscr{L}^\psi(R, A \cup \Pi) \subseteq \mathscr{L}^\psi(R, A)$. By assumptions of our theorem, $h^e(\Pi) \subseteq h^e(Y) \subseteq Cn(R, A)$ and $h^e(\psi) \notin Cn(R, A)$. The rule r is structural, so we get $\langle h^e(\Pi), h^e(\psi) \rangle \in r$. Thus, $r \notin \mathrm{Adm}(R, A)$. $\square$

Corollary 3.69 (∞). *If* $\langle R, A \rangle \in \mathrm{Inv}$, *then*

$$\langle R, A \rangle \in \mathrm{SCpl} \quad \Rightarrow \quad \forall_{Y \in \mathscr{L}(R,A)} \; \exists_{e \colon At \to S} \; Y = h^{e^{-1}}\big(Cn(R, A)\big).$$

This follows directly from Theorem 3.68; the assumption that Cn_{RA} is finitistic has not been used in the proof of the implication ($\Rightarrow$). A counterexample below shows that the reverse implication does not hold:

Example. Let $\mathscr{S} = \langle S_{FG}; F, G \rangle$ where F, G are unary connectives. We define two rules in this language: $r_1 = \{\langle \alpha, F\alpha \rangle \,:\, \alpha \in S_{FG}\}$, $r_2 = \{\langle \alpha, G\alpha \rangle \,:\, \alpha \in S_{FG}\}$, $A = Sb(\{FFp, FGp, GFp, GGp\})$, $R = \{r_1, r_2\}$. We have now $\langle R, A \rangle \in \mathrm{Inv}$ and

    1) $p, Gp, Fp \notin Cn(R, A)$,

2) $Cn(R, A \cup \{p\}) = S_{FG}$,

3) $Cn(R, S_{FG} \setminus At) = S_{FG} \setminus At$.

Thus, the only complete overset of $Cn(R, A)$ is $Y = S_{FG} \setminus At$. For $e(p) = Fp$ we have $h^{e^{-1}}(A) = Y$. Define now $r_3 = \{\langle F\alpha, G\alpha \rangle : \alpha \in S_{FG}\}$; it is easily seen that $r \in \text{Struct} \cap \text{Adm}(R, A) \setminus \text{Der}(R, A)$. Thus $\langle R, A \rangle \notin \text{SCpl}$.

Let $\mathfrak{M}, \mathfrak{N}$ be two logical matrices. We say that $\mathfrak{M}$ is *embeddable* in $\mathfrak{N}$ if $\mathfrak{M}$ is isomorphic with some submatrix of $\mathfrak{N}$. We confine our investigations now to logical systems with the equivalence connective. That is, we assume that

$$\alpha \sim_{R, A \cup Y} \beta \quad \text{iff} \quad \alpha \to \beta, \beta \to \alpha \in Cn(R, A \cup Y)$$

is a congruence in the Lindenbaum matrix $\mathfrak{M}^{R, A \cup Y}$ for each $Y \subseteq S$. We will write $\sim_Y$ instead of $\sim_{R, A \cup Y}$ if the system $\langle R, A \rangle$ is fixed. We get the following characterization of structural completeness:

Theorem 3.70 (∞). *If $\langle R, A \rangle$ is an invariant system with equivalence and Cn_{RA} is finitistic, then the following conditions are equivalent:*

(i) $\langle R, A \rangle \in \text{SCpl}$;

(ii) $\mathfrak{M}^{R, A}/ \sim_Y$ *is embeddable in* $\mathfrak{M}^{R, A}/ \sim_A$ *for each* $\psi \in S$ *and* $Y \in \mathscr{L}^\psi(R, A)$;

(iii) $\overrightarrow{\mathscr{M}^{R, A}} \leqslant \overrightarrow{\mathscr{M}^{R, Y}}$ *for each* $\psi \in S$ *and* $Y \in \mathscr{L}^\psi(R, A)$.

Proof. (i)$\Rightarrow$(ii): If $Y \in \mathscr{L}^\psi(R, A)$, then (by 3.68) there is an endomorphism $h \colon S \to S$ such that $Y = h^{-1}\big(Cn(R, A)\big)$. Define $f \colon S/\sim_y \to S/\sim_A$ by

$$f([\alpha]_Y) = [h(\alpha)]_A, \quad \text{for every } \alpha \in S.$$

It can be verified that f is a one-to-one homomorphism and

$$[\alpha]_Y \in Cn(R, A \cup Y)/\sim_Y \quad \Leftrightarrow \quad f([\alpha]_Y) \in Cn(R, A)/\sim_A.$$

(ii)$\Rightarrow$(i): This follows immediately from Corollary 2.37.

(iii)$\Rightarrow$(i): Assume that $\psi \notin Cn(R, A \cup X)$ for some $X \subseteq S$ and $\psi \in S$. Then, by Theorem 3.26, there exists a set $Y \in \mathscr{L}^\psi(R, A \cup X)$. Let us consider the Lindenbaum matrix $\mathfrak{M}^{R, Y}$. Obviously, we have $\psi \notin \overrightarrow{\mathfrak{M}^{R, Y}}(X)$ and hence by (iii) $\psi \notin \overrightarrow{\mathfrak{M}^{R, A}}(X)$. Thus, it has been proved that $\overrightarrow{\mathfrak{M}^{R, A}}(X) \subseteq Cn(R, A \cup X)$ for each $X \subseteq S$. According to Corollary 3.66, we get $\langle R, A \rangle \in \text{SCpl}$. $\square$

Let us notice that our assumption that $\langle R, A \rangle$ is a system with equivalence is not necessary for the proof of the equivalence (i)$\Leftrightarrow$(iii) but it is required for (i)$\Leftrightarrow$(ii). The later equivalence has been proved in [105], 1977.

We will now give a short survey of the most important results concerning structural completeness of some concrete propositional logics.

Classical logic

The invariant version of classical propositional logic, although Post-incomplete, is structurally complete (see [79], 1971):

Theorem 3.71. $\langle R_0, Sb(A_2) \rangle \in \mathrm{SCpl}$.

Proof. Let us assume that $\alpha \notin Cn\big(R_0, Sb(A_2) \cup X\big)$. By the generalized completeness theorem 3.38, there exists a valuation $v \colon At \to \{0,1\}$ such that $h^v(X) \subseteq \{1\}$ and $h^v(\alpha) = 0$ (where $h^v \colon S_2 \to \{0,1\}$). Let us define $e \colon At \to S_2$ as follows:

$$e(\gamma) = \begin{cases} \gamma \to \gamma, & \text{if } v(\gamma) = 1 \\ \sim (\gamma \to \gamma), & \text{if } v(\gamma) = 0. \end{cases}$$

Obviously, we have $h^w \circ e = v$ for every valuation $w \colon At \to \{0,1\}$. Then we get $h^e(X) \subseteq E(\mathfrak{M}_2) = Cn\big(R_0, Sb(A_2)\big)$ and $h^e(\alpha) \notin Cn\big(R_0, Sb(A_2)\big)$. Consequently, $\langle X, \alpha \rangle \notin r_{Cn(R_0, Sb(A_2))}$. Thus, we have proved $r_{Cn(R_0, Sb(A_2))} \in \mathrm{Der}\big(R_0, Sb(A_2)\big)$ and hence, by Corollary 3.64, $\langle R_0, Sb(A_2) \rangle \in \mathrm{SCpl}$. □

It seems worth noticing that Theorem 3.71 can also be proved without using the completeness theorem (and without using Post's theorem); the structural completeness easily follows from Lindenbaum–Łoś's theorem.

Classical logic is very often understood as a system $\langle R, A \rangle$ such that $Cn(R, A) = Z_2$ (it should be emphasized that we do not accept such a definition). Accepting this definition one easily proves there are 'classical' systems which are not structurally complete:

Example. Let $\bar{r}_o = \{\langle\{\alpha, \alpha \to \beta\}, \beta\rangle : \alpha \in S_2, \beta \in S_2 \setminus At\}$ and $\overline{R}_{0*} = \{\bar{r}_0, r_*\}$. Notice that the rule r_0 is admissible in the system $\langle \overline{R}_{0*}, A_2 \rangle$ (and consequently $Cn(\overline{R}_{0*}, A_2) = Z_2$) but $r_0 \notin \mathrm{Der}(\overline{R}_{0*}, A_2)$ since $Cn(\overline{R}_{0*}, S_2 \setminus At) = S_2 \setminus At$. Thus, we have $\langle \overline{R}_{0*}, A_2 \rangle \notin \mathrm{SCpl}$.

Intermediate logics

Structural incompleteness of intuitionistic propositional logic, defined in the language with $\{\to, \cdot, +, \sim\}$, has been already proved by Minc [71], 1972.

Theorem 3.72. $\langle R_{0*}, A_i \rangle \notin \mathrm{SCpl}$.

The following structural rules, for instance, are admissible in intuitionistic logic but they are not derivable there:

$$\frac{\sim \alpha \to \beta + \gamma}{(\sim \alpha \to \beta) + (\sim \alpha \to \gamma)}, \qquad \frac{(\alpha \to \beta) \to \alpha + \gamma}{((\alpha \to \beta) \to \alpha) + ((\alpha \to \beta) \to \gamma)}.$$

Some fragments of the intuitionistic logic may be, however, structurally complete. There are many results on this subject proved by A. Wroński and T. Prucnal; some of them may be found in [58], 1983. For instance, we have

Theorem 3.73. $\langle R_0, Sb(A_i^{\rightarrow \cdots \sim}) \rangle \in$ SCpl.

The rule basis for the structural complete strengthening of $\langle R_0, Sb(A_i) \rangle$, see Corollary 3.62, has been given by R. Iemhoff [43], 2001. It should also be mentioned that not only the Lindenbaum matrix is not strongly adequate for the intuitionitsic logic $\langle R_0, Sb(A_i) \rangle$, see Corollary 3.67, but there does not exist any countable matrix strongly adequate for this logic, see [151], 1974). The logic $\langle R_0, Sb(A_i) \rangle$ does possess a strongly adequate matrix of cardinality $\mathfrak{c}$.

There is also an extensive literature on structural completeness of intermediate logics. Let us mention, for instance, Citkin [12], 1977 and Rybakov [112], 1997. As concerns this subject we only quote without proof (see Dzik [18], 1973:

Theorem 3.74.

(i) $\langle R_0, Sb(H_n) \rangle \in$ SCpl,   *for each* $n \geqslant 2$;

(ii) $\langle R_{0*}, A_l \rangle \in$ SCpl;

(iii) (∞) $\langle R_0, Sb(A_l) \rangle \notin$ SCpl;

(iv) (fin) $\langle R_0, Sb(A_l) \rangle \in$ SCpl.

Many-valued logics of Łukasiewicz

Let $\mathfrak{M}_n$ be the n-valued Łukasiewicz's matrix $\langle \langle A_n, f_n^{\rightarrow}, f_n^+, f_n^{\cdot}, f_n^{\sim} \rangle, \{1\} \rangle$. We shall prove now the fundamental lemma on the structural completeness of finite-valued Łukasiewicz logics.

Lemma 3.75. $\overrightarrow{\mathfrak{M}_n \times \mathfrak{M}_2} \in$ SCPL,   *for every* $n \geqslant 2$.

Proof. Suppose that $n \geqslant 2$ is a fixed natural number. For every $x \in A_n$ (the universe of $\mathfrak{M}_n$) let α_x^0, α_x^1 be two formulas defined as follows (we use the abbreviations: $p \rightarrow^0 q = q$, $p \rightarrow^{k+1} q = p \rightarrow (p \rightarrow^k q)$):

$$\alpha_1^1 = p \rightarrow p,$$
$$\alpha_{\frac{n-2}{n-1}}^1 = \left((p \rightarrow^{n-2} q) \rightarrow p \right) \rightarrow p,$$
$$\alpha_0^1 = \left((p \rightarrow^{n-2} q) \rightarrow^{n-1} q \right) \rightarrow \left((p \rightarrow^{n-1} q) \rightarrow q \right), \quad \text{if } n > 2,$$
$$\alpha_x^1 = \alpha_{\frac{n-2}{n-1}}^1 \rightarrow^{x \cdot (n-1)} \alpha_0^1, \quad \text{if } 0 \leqslant x < \frac{n-2}{n-1},$$
$$\alpha_x^0 = \sim \alpha_{1-x}^1.$$

It can be verified that for every $v \colon At \rightarrow A_n$,

$$h^v(\alpha_x^i) = \begin{cases} i & \text{if } v(p) \neq \frac{n-2}{n-1} \text{ or } v(q) \neq 0 \\ x & \text{if } v(p) = \frac{n-2}{n-1} \text{ and } v(q) = 0. \end{cases}$$

Let us define a relation $\approx_n$ on S_2: $\alpha \approx_n \beta$ iff $\alpha \equiv \beta \in E(\mathfrak{M}_n)$. Of course, $\approx_n$ is a congruence-relation on the Lindenbaum matrix $\mathfrak{M}^{R_0, Sb(\text{Ł}_n)}$ and $\alpha \approx_n \beta$ iff $h^v(\alpha) = h^v(\beta)$ for every $v \colon At \to A_n$. Let $f \colon A_n \times A_2 \to S_2/\approx_n$ be defined as follows $f(\langle x, i \rangle) = |\alpha_x^i|$. One can easily prove that f is a homomorphism from $\mathfrak{M}_n \times \mathfrak{M}_2$ into $S_2/\approx_n$. Moreover, f is one-to-one and hence $\mathfrak{M}_n \times \mathfrak{M}_2$ is isomorphic with some submatrix of the Lindenbaum matrix $\mathfrak{M}^{R_0, Sb(\text{Ł}_n)}/\approx_n$. Thus, we have $\overrightarrow{\mathfrak{M}^{R_0, Sb(\text{Ł}_n)}} \leqslant \overrightarrow{\mathfrak{M}_n \times \mathfrak{M}_2}$ by Corollary 2.37.

On the other hand, since $\overrightarrow{\mathfrak{M}_n}(\emptyset) = \overrightarrow{\mathfrak{M}^{R_0, Sb(\text{Ł}_n)}}(\emptyset) = \overrightarrow{\mathfrak{M}_n \times \mathfrak{M}_2}(\emptyset)$ we infer from Makinson's theorem 3.59 that $\overrightarrow{\mathfrak{M}_n \times \mathfrak{M}_2} \leqslant \overrightarrow{\mathfrak{M}^{R_0, Sb(\text{Ł}_n)}}$. Hence, we get $\overrightarrow{\mathfrak{M}_n \times \mathfrak{M}_2} = \overrightarrow{\mathfrak{M}^{R_0, Sb(\text{Ł}_n)}} \in \text{SCPL}$.   □

Thus, from Theorem 2.41 it follows that

$$\overrightarrow{\mathfrak{M}_n \times \mathfrak{M}_2}(X) = \begin{cases} \overrightarrow{\mathfrak{M}_n}(X) & \text{if } X \in \text{Sat}(\mathfrak{M}_2) \\ S_2 & \text{if } X \notin \text{Sat}(\mathfrak{M}_2). \end{cases}$$

It it easy to see that $\overrightarrow{\mathfrak{M}_n \times \mathfrak{M}_2}(X) > \overrightarrow{\mathfrak{M}_n}$ if $n \neq 2$ (for example, if one takes $\alpha = [(\sim p \to p) \overset{n-1}{\longrightarrow} p] \to \sim ((\sim p \to p) \to p)$, then $\overrightarrow{\mathfrak{M}_n}(\alpha) \neq S_2$ and $\alpha \notin \text{Sat}(\mathfrak{M}_2)$). Thus as an immediate result of 3.59 we infer (see [79], 1971 and [122], 1972)

Corollary 3.76. $\langle R_0, Sb(\text{Ł}_n) \rangle \notin \text{SCpl}$, *for every $n > 2$.*

We shall now prove the theorem on structural completeness of Łukasiewicz's propositional logic:

Theorem 3.77. $\langle R_{0*}, \text{Ł}_n \rangle \in \text{SCpl}$, *for every $n \geqslant 2$.*

Proof. To prove that $\overrightarrow{\mathfrak{M}_n}(Sb(X)) = \overrightarrow{\mathfrak{M}_n \times \mathfrak{M}_2}(Sb(X))$ for every $X \subseteq S_2$, it suffices to show (see 2.41) that $\overrightarrow{\mathfrak{M}_n}(Sb(X)) = S_2$ if $Sb(X) \notin \text{Sat}(\mathfrak{M}_2)$. Suppose that it is not true, i.e., $Sb(X) \notin \text{Sat}(\mathfrak{M}_2)$ and $\overrightarrow{\mathfrak{M}_n}(Sb(X)) \neq S_2$ for some $X \subseteq S_2$. Then there exists a valuation $v \colon At \to A_n$ such that $h^v(Sb(X)) \subseteq \{1\}$. Let $e \colon At \to S_2$ be a substitution defined as follows: $e(\gamma) = \gamma \to \gamma$ for each $\gamma \in At$. We have $w = h^v \circ e \colon At \to \{0, 1\}$, so $h^w(Sb(X)) = h^v h^e(Sb(X)) \subseteq h^v(Sb(X)) \subseteq \{1\}$ which is a contradiction. Hence, for each X, we get $\overrightarrow{\mathfrak{M}_n \times \mathfrak{M}_2}(Sb(X)) = \overrightarrow{\mathfrak{M}_n}(Sb(X)) = Cn(R_0, Sb(\text{Ł}_n) \cup Sb(X)) = Cn(R_{0*}, \text{Ł}_n \cup X)$. Thus, $\langle R_{0*}, \text{Ł}_n \rangle \in \text{SCpl}$.   □

This theorem was proved first for $n = 2, 3$ in [79], 1971 and later generalized in [122], 1972. The above proof was given in [130], 1976.

We complete the examination of many-valued Łukasiewicz logics by the following results:

Theorem 3.78. $\langle R_0, Sb(\overrightarrow{\text{Ł}_n^{+\cdot}}) \rangle \in \text{SCpl}$, *for each $n \geqslant 2$.*

Theorem 3.79. $\langle R_{0*}, \overrightarrow{\text{Ł}_\infty^{+\cdot}} \rangle \notin \text{SCpl}$.

The structural completeness of the pure implicational logics of Łukasiewicz has been proved by T. Prucnal to whom belongs also 3.79 (unpublished). The proof of 3.79 can be found in [131], 1978.

Modal logics

In this part we consider the problem of structural completeness of Lewis's modal system $S5$.

Theorem 3.80 (fin).

   (i) $\langle R_{0*a}, A_{S5} \rangle \in$ SCpl;

   (ii) $\langle R_{0a}, Sb(A_{S5}) \rangle \notin$ SCpl.

Proof. Assume the abbreviation $Z_5 = Cn(R_{0*a}, A_{S5})$ and define

$$\alpha \approx \beta \quad \text{iff} \quad \alpha \equiv \beta \in Z_5, \qquad \text{for} \quad \alpha, \beta \in S_2.$$

It is easy to observe that $\approx$ is a congruence relation on $\mathscr{S}_2$. Let $\phi \in S_2$. Define $e_\phi \colon At \to S_2$ by $e_\phi(\gamma) = \phi \cdot \gamma$ for each $\gamma \in At$. We write e_0 for $\phi =\sim (p_0 \to p_0)$ and e_1 if $\phi = p_0 \to p_0$. By (33) and (34) (see Chapter 1) we have

(A) $h^{e_1}(\alpha) \approx \alpha$,   for every $\alpha \in S_2$,

(B) $h^{e_0}(\alpha) \in Z_5$ or $\sim h^{e_0}(\alpha) \in Z_5$,   for every $\alpha \in S_2$

(the easy proof of (B), by induction on the length of α, is left to the reader).

(C) $(\phi \equiv \psi) \to \left(h^{e_\phi}(\alpha) \equiv h^{e_\psi}(\alpha) \right) \in Z_5$,   for each $\phi, \psi, \alpha \in S_2$.

Indeed, we get $\phi \cdot \gamma \equiv \psi \cdot \gamma \in Cn(R_{0*a}, A_{S5} \cup \{\phi \equiv \psi\})$ for $\gamma \in At$ and hence from 1.72 we get $h^{e_\phi}(\alpha) \equiv h^{e_\psi}(\alpha) \in Cn(R_{0*a}, A_{S5} \cup \{\phi \equiv \psi\})$ which gives (C). As an immediate corollary of (C) and (A) we obtain

(D) $\left(\phi \equiv (p_0 \to p) \right) \to \left(h^{e_\phi}(\alpha) \equiv \alpha \right) \in Z_5$,
   $\left(\phi \equiv\sim (p_0 \to p) \right) \to \left(h^{e_\phi}(\alpha) \equiv h^{e_\psi}(\alpha) \right) \in Z_5$,   for each $\phi, \psi \in S_2$.

We have then

$$\Box\phi \to \left(h^{e_\phi}(\alpha) \equiv \alpha \right) \in Z_5 \text{ and}$$
$$\Box\sim\phi \to \left(h^{e_\phi}(\alpha) \equiv h^{e_0}(\alpha) \right) \in Z_5, \quad \text{for each } \alpha, \phi \in S_2.$$

Assume that $\phi \in S_L$ (then also $\sim \phi \in S_\Box$ — see 1.73), hence $\phi \approx \Box\phi$ and $\Box\sim\phi \approx\sim\phi$. Thus,

$$\phi \to \left(h^{e_\phi}(\alpha) \equiv \alpha \right) \in Z_5 \text{ and}$$
$$\sim\phi \to \left(h^{e_\phi}(\alpha) \equiv h^{e_0}(\alpha) \right) \in Z_5 \quad \text{for each } \phi \in S_\Box, \alpha \in S_2.$$

Using (35) and (40) we get

$$\phi \cdot h^{e_\phi}(\alpha) +\sim \phi \cdot h^{e_\phi}(\alpha) \approx \phi \cdot \alpha +\sim \phi \cdot h^{e_0}(\alpha)$$

and hence by (30), (32)

(E) $h^{e\phi}(\alpha) \approx \phi \cdot \alpha + \sim \phi \cdot h^{e_0}(\alpha)$ for each $\phi \in S_\square$, $\alpha \in S_2$.

Assume that for some $\alpha_1, \ldots, \alpha_k, \beta \in S_2$ we have

(F) $h^e(\{\alpha_1, \ldots, \alpha_k\}) \subseteq Z_5 \quad \Rightarrow \quad h^e(\beta) \in Z_5$,   for every $e\colon At \to S_2$.

We have to show that $\beta \in Cn(R_{0*a}, A_{S5} \cup \{\alpha_1, \ldots, \alpha_k\})$. Let $\alpha = \alpha_1 \cdot \ldots \cdot \alpha_k$ and suppose that $\sim h^{e_0}(\alpha) \in Z_5$. Then $S_2 = Cn(R_{0*a}, A_{S5} \cup \{\alpha\})$ and hence $\beta \in Cn(R_{0*a}, A_{S5} \cup \{\alpha_1, \ldots, \alpha_k\})$. Thus, by (B), we can assume that $h^{e_0}(\alpha) \in Z_5$ and hence by (E),

$$h^{e\square\alpha}(\alpha) \approx \square\alpha + \sim L\alpha.$$

It follows now from (F) that $h^{e\square\alpha}(\beta) \in Z_5$ and hence $\square\alpha \cdot \beta + \sim \square\alpha \cdot h^{e_0}(\beta) \in Z_5$. Thus, by means or r_a and by (30), we get

$$\beta \in Cn(R_{0*a}, A_{S5} \cup \{\square\alpha\})$$

and since Gödel's rule is derivable in the system $\langle R_{0*a}, A_{S5} \rangle$ (see 1.75) we get $\beta \in Cn(R_{0*a}, A_{S5} \cup \{\alpha\}) = Cn(R_{0*a}, A_{S5} \cup \{\alpha_1, \ldots, \alpha_k\})$. Gödel's rule $r_\square$ is structural and admissible in $\langle R_{0a}, Sb(A_{S5}) \rangle$ but $r_\square$ is not derivable in this system: the reverse assumption leads to the conclusion that $p \to ((p \to p) \to p) \in Z_5$, which is false. $\square$

The proof of the above theorem is in [99], 1972. Note, however that the main result of [99] is incorrect as stated: the author did not take into account that the systems $\langle R_{0*}, A_{S5} \rangle$ and $\langle R_{0*a}, A_{S5} \rangle$ are non-equivalent. Let us remark that non-derivability of $r_\square$ (or r_a) in the system $\langle R_{0*}, A_{S5} \rangle$ yields a structural incompleteness of the system $\langle R_{0*} A_{S5} \rangle$.

From Biela's paper [3], 1975 it follows that all subsystems of $S3.5$ are structurally incomplete. Moreover, it is known (but unpublished) that both versions of $S4$ (Gödel's and Meredith's) are structurally incomplete, too.

3.5   Some related concepts

Many authors restrict propositional logic to invariant propositional systems only. This approach gives some advantages. It should be noted, however, that this also means reduction of the deductive power of propositional systems (by eliminating the substitution rule from the set of derivable rules!).

The observation that the notion of Post-completeness does not meet the case of invariant logics forced logicians to introduce other concepts of completeness for these systems. The notion of structural completeness (if restricted to invariant logics) can serve as an example. Another example can be given by assuming $\Gamma = S$ in Definition 3.1 (ii). It is easy to see that $\langle R, A \rangle \in S\text{-Max}$ if and only if $\langle R, A \rangle$, if consistent, is a maximal element in the preordered set of all invariant consistent logics. The third concept of this kind is the notion of saturation introduced by Wójcicki (cf. [123] 1973):

Definition 3.81. A propositional logic $\langle R, X \rangle \in \text{Inv}$ is said to be *saturated*, formally $\langle R, X \rangle \in \text{Std}$, iff $\text{Der}(R \cup \{r_*\}, X) \cap \text{Struct} \subseteq \text{Der}(R, X)$.

It is easily seen that this notion matches invariant logics, as for systems $\langle R, X \rangle$ such that $r_* \in \text{Der}(R, X)$ the definitional inclusion is trivially fulfilled. If $\langle R, X \rangle$ is saturated, then the structural rules derivable in this logic are the same as the structural rules derivable in $\langle R \cup \{r_*\}, X \rangle$. Let us try to establish some connections between the three above concepts.

Lemma 3.82. $S\text{-Max} \subseteq \text{SCpl} \cap \text{Inv} \subseteq \text{Std}$.

This follows immediately from the definitions and Lemma 3.2. It has already been shown (see the example on page 92) that S-Max is a proper subclass of $\text{SCpl} \cap \text{Inv}$. To complete our discussion let us prove that there is a saturated logic which is not structurally complete:

Example. Let us consider the system $\langle \overline{R}_{0*}, A_2 \rangle$ from the example on page 116, where it is proved that this logic is not structurally complete. By the deduction theorem for the system $\langle R_0, Sb(A_2) \rangle$ and by the equality $Cn(R_0, Sb(A_2)) = Cn(\overline{R}_0, Sb(A_2))$, where $\overline{R}_0 = \{\overline{r}_0\}$, we get

$$(\star) \qquad \alpha \in Cn(R_0, Sb(A_2) \cup X) \quad \Rightarrow \quad \alpha \in At \ \lor \ \alpha \in Cn(\overline{R}_0, Sb(A_2) \cup X).$$

Since the set $S_2 \setminus At$ is closed under the rule $\overline{r}_0$, it follows from the above statement and from the fact that $\langle R_{0*}, A_2 \rangle$ is Post-complete that

$$Cn(\overline{R}_{0*}, A_2 \cup X) = \begin{cases} Z_2 & \text{if } X \subseteq Z_2 \\ S_2 & \text{if } X \cap At \neq \emptyset \\ S_2 \setminus At & \text{otherwise.} \end{cases}$$

Let r be a structural rule which is not derivable in the system $\langle \overline{R}_0, Sb(A_2) \rangle$ that is $r \in \text{Struct} \setminus \text{Der}(\overline{R}_0, Sb(A_2))$. So $\phi \notin Cn(\overline{R}_0, Sb(A_2) \cup \Pi)$ for some $\langle \Pi, \phi \rangle \in r$. We have to prove that $r \notin \text{Der}(\overline{R}_{0*}, Sb(A_2))$. If $\phi \notin Cn(R_0, Sb(A_2) \cup \Pi)$, then $r \notin \text{Adm}(R_0, Sb(A_2)) = \text{Adm}(R_{0*}, A_2)$ by the structural completeness of the classical logic. But $Cn(\overline{R}_{0*}, A_2) = Cn(R_{0*}, A_2)$ and hence $r \notin \text{Adm}(\overline{R}_*, A_2) \supseteq \text{Der}(\overline{R}_{0*}, A_2) = \text{Der}(\overline{R}_{0*}, Sb(A_2))$. So, we can assume $\phi \in Cn(R_0, Sb(A_2) \cup \Pi)$. It follows from $(\star)$ that $\phi \in At$. Consider a substitution $e \colon At \to S_2$ such that $e(\phi) = \phi$ and $e(\gamma) \notin At$ for every $\gamma \in At \setminus \{\gamma\}$. Obviously, $\phi \notin \Pi$ because $\phi \notin Cn(\overline{R}_0, Sb(A_2) \cup \Pi)$. Therefore, $h^e(\Pi) \subseteq S \setminus At$ and $h^e(\phi) = \phi \in At$. Thus, we get $h^e(\phi) \notin Cn(\overline{R}_{0*}, Sb(A_2) \cup h^e(\Pi))$. Since $\langle \Pi, \phi \rangle \in r$ and r is a structural rule, we have $\langle h^e(\Pi), h^e(\phi) \rangle \in r$ and consequently $r \notin \text{Der}(\overline{R}_{0*}, A_2)$.

Another concept of completeness (we will call it *pseudo-completeness*) for invariant logics was introduced by [123], 1973.

Definition 3.83. Let $\langle R, A \rangle \in \mathrm{Inv}$, then

$$\langle R, A \rangle \in p\text{-Cpl} \quad \Leftrightarrow \quad \langle R \cup \{r_*\}, A \rangle \in \mathrm{Cpl}.$$

It is easy to see that the above notion is obtained by a simple translation of the notion of Post-completeness into the class Inv. Thus, we get

Lemma 3.84. *If* $\langle R, A \rangle \in \mathrm{Inv}$, *then*

$$\langle R, A \rangle \in p\text{-Cpl} \quad \Leftrightarrow \quad Cn\big(R, A \cup Sb(\{\alpha\})\big) = S \ \text{ for every } \ \alpha \notin Cn(R, A).$$

Maximality in the family of consistent invariant logics is a very strong property. Among the concepts considered here, it appears to be the best counterpart — for invariant logics — of the notion of Post-completeness (see [80], 1974). By Lemmas 3.3 and 3.11, we easily get

Corollary 3.85. *Let* $\langle R, A \rangle \in \mathrm{Inv}$. *Then*

(i) $\langle R, A \rangle \in S\text{-Max} \Leftrightarrow \forall_{r \in \mathrm{Struct}} \big(\langle R \cup \{r\}, A \rangle \in \mathrm{Cns} \Rightarrow r \in \mathrm{Der}(R, A) \big);$

(ii) $\langle R, A \rangle \in p\text{-Cpl} \Leftrightarrow \forall_{r \in \mathrm{Struct}} \big(\langle R \cup \{r\}, A \rangle \in \mathrm{Cns} \Rightarrow r \in \mathrm{Der}(R \cup \{r_*\}, A) \big).$

One can easily prove, see [123], 1973:

Theorem 3.86. *If* $\langle R, A \rangle \in \mathrm{Inv}$, *then the following conditions are equivalent:*

(i) $\langle R, A \rangle \in S\text{-Max};$

(ii) $\langle R, A \rangle \in \mathrm{SCpl} \cap p\text{-Cpl};$

(iii) $\langle R, A \rangle \in \mathrm{Std} \cap p\text{-Cpl}.$

Moreover, on the basis of Theorem 3.14 we obtain:

Corollary 3.87 (fin). *For every* $\langle R, A \rangle \in \mathrm{Inv} \cap \mathrm{Cns}$ *there is* $\langle R_1, A_1 \rangle \in S\text{-Max}$ *such that* $\langle R, A \rangle \preccurlyeq \langle R_1, A_1 \rangle \in \mathrm{Cns}.$

In other words any structural, consistent and finitistic consequence operation can be extended to a maximal and consistent one.

Corollary 3.87 (∞). *For every* $\langle R, A \rangle \in \mathrm{Inv} \cap \mathrm{Cns} \cap \mathrm{Comp}$ *there is* $\langle R_1, A_1 \rangle \in S\text{-Max such that}$ $\langle R, A \rangle \preccurlyeq \langle R_1, A_1 \rangle \in \mathrm{Cns}.$

The example on page 97 shows that the assumption $\langle R, A \rangle \in \mathrm{Comp}$ cannot be omitted. By Theorems 3.86, 3.71 and 3.39 we get (see [75], 1968)

Theorem 3.88. $\langle R_0, Sb(A_2) \rangle \in S\text{-Max}.$

All other invariant logics considered in this book are not S-maximal. It is because they are weaker than the classical system $\langle R_0, Sb(A_2) \rangle$. Theorem 3.86 allows us to establish other connections between S-maximality and saturation ([123], 1973).

Corollary 3.89. *If* $\langle R, X \rangle \in \mathrm{Inv} \cap p\text{-Cpl}$, *then*

$$\langle R, X \rangle \in \mathrm{SCpl} \qquad \Leftrightarrow \qquad \langle R, X \rangle \in S\text{-Max}.$$

The next few results concern saturation and we prove that this notion is a derivative of structural completeness.

Lemma 3.90. *Let* $\langle R, A \rangle \in \mathrm{Inv}$. *Then* $\langle R, A \rangle \in \mathrm{Std}$ *if and only if*

$$\big(\langle R_1 \cup \{r_*\}, A_1 \rangle \preccurlyeq \langle R \cup \{r_*\}, A \rangle \Rightarrow \langle R_1, A_1 \rangle \preccurlyeq \langle R, A \rangle, \text{ for every } \langle R_1, A_1 \rangle \in \mathrm{Inv} \big).$$

Proof. ($\Rightarrow$): Let $\langle R_1 \cup \{r_*\}, A_1 \rangle \preccurlyeq \langle R \cup \{r_*\}, A \rangle$ for some invariant $\langle R_1, A_1 \rangle$ and let $\langle R, A \rangle \in \mathrm{Std}$. Then $R_1 \subseteq \mathrm{Struct} \cap \mathrm{Der}(R \cup \{r\}, A) \subseteq \mathrm{Der}(R, A)$ and $A_1 \subseteq Cn(R \cup \{r_*\}, A) = Cn(R, A)$. Thus, $\langle R_1, A_1 \rangle \preccurlyeq \langle R, A \rangle$.

($\Leftarrow$): From the assumptions we get $\langle \mathrm{Der}(R \cup \{r_*\}, A) \cap \mathrm{Struct}, A \rangle \preccurlyeq \langle R, A \rangle$ and hence $\mathrm{Der}(R \cup \{r_*\}, A) \cap \mathrm{Struct} \subseteq \mathrm{Der}(R, A)$. Thus, we get $\langle R, A \rangle \in \mathrm{Std}$. $\square$

Adopting the above theorem to the consequence formalism we obtain

Corollary 3.90 (∞). *If* $Cn \in \mathrm{STRUCT}$, *then*

$$Cn \in \mathrm{STD} \quad \Leftrightarrow \quad \forall_{Cn_1 \in \mathrm{STRUCT}} \big(Cn_1 \circ Sb \leqslant Cn \circ Sb \Rightarrow Cn_1 \leqslant Cn \big).$$

It is also clear that

Corollary 3.91. *If* $\langle R, A \rangle \in \mathrm{Inv}$ *and* $\langle R \cup \{r_*\}, A \rangle \in \mathrm{SCpl}$, *then*

$$\langle R, A \rangle \in \mathrm{Std} \qquad \Leftrightarrow \qquad \langle R, A \rangle \in \mathrm{SCpl}.$$

The above statement is easy to anticipate and it can be easily deduced from the definitions of the involved notions. More surprising connections between saturation and structural completeness are given below. We recall that, for any family $\{\langle R_t, A_t \rangle : t \in T\}$ of propositional systems, the symbol $\prod_{t \in T} \langle R_t, A_t \rangle$ stands for the system $\langle \bigcap_{t \in T} \mathrm{Der}(R_t, A_t), \bigcap_{t \in T} Cn(R_t, A_t) \rangle$.

Lemma 3.92. *For every family* $\{\langle R_t, A_t \rangle : t \in T\} \subseteq \mathrm{Std} \cap \mathrm{Inv}$, *we have*

$$\prod_{t \in T} \langle R_t, A_t \rangle \in \mathrm{Std}.$$

Proof. Let $\{\langle R_t, A_t \rangle : t \in T\} \subseteq \mathrm{Inv}$. We know that (see 1.61) there exists an invariant logic $\langle R, A \rangle$ such that $\langle R, A \rangle \approx \prod_{t \in T} \langle R_t, A_t \rangle$. Let us assume now that $\{\langle R_t, A_t \rangle : t \in T\} \subseteq \mathrm{Std}$ and suppose that $r \in \mathrm{Der}(R \cup \{r_*\}, A) \cap \mathrm{Struct}$. Then, for every $\langle \Pi, \alpha \rangle \in r$, we have

$$\alpha \in Cn(R \cup \{r_*\}, A \cup \Pi) = Cn(R, A \cup Sb(\Pi)) = Cn(R, A \cup Sb(\Pi))$$
$$\subseteq \bigcap_{t \in T} Cn(R_t, A_t \cup Sb(\Pi)) = \bigcap_{t \in T} Cn(R \cup \{r_*\}, A_t \cup \Pi).$$

Thus, we have $r \in \text{Der}(R_t \cup \{r_*\}, A_t) \cap \text{Struct}$ for each $t \in T$. By the saturation of the system $\langle R_t, A_t \rangle$, we get $r \in \text{Der}(R_t, A_t)$ for each $t \in T$ and hence (see Lemma 1.52) $r \in \bigcap_{t \in T} \text{Der}(R_t, A_t) = \text{Der}\left(\prod_{t \in T}(R_t, A_t)\right) = \text{Der}(R, A)$. $\qquad\square$

Theorem 3.93. *If* $\langle R, A \rangle \in \text{Inv}$, *then* $\langle R, A \rangle \in \text{Std}$ *if and only if*

$$\prod_{t \in T} \langle R_t, A_t \rangle \approx \langle R, A \rangle \quad \textit{for some family} \ \{\langle R_t, A_t \rangle : t \in T\} \subseteq \text{Inv} \cap \text{SCpl}.$$

Proof. The implication ($\Leftarrow$) follows directly from 3.92 and 3.82.

($\Rightarrow$): Let $R_X = \text{Adm}(R, A \cup Sb(X)) \cap \text{Struct}$ for any $X \subseteq S$. Obviously, $\text{Adm}(R_X, A \cup Sb(X)) \cap \text{Struct} = \text{Adm}(\text{Adm}(R, A \cup Sb(X)), A \cup Sb(X)) \cap \text{Struct} = R_X$ and hence $\langle R_X, A \cup Sb(X) \rangle \in \text{Inv} \cap \text{SCpl}$ for any $X \subseteq S$. Moreover, $R \subseteq R_X$ for each $X \subseteq S$. Thus, $\langle R, A \rangle \preccurlyeq \prod_{X \subseteq S} \langle R_X, A \cup Sb(X) \rangle$. We have to prove that $\prod_{X \subseteq S} \langle R_X, A \cup Sb(X) \rangle \preccurlyeq \langle R, A \rangle$, that is (see 1.51)

$$\bigcap_{X \subseteq S} Cn(R_X, A \cup Sb(X) \cup \Pi) \subseteq Cn(R, A \cup \Pi), \quad \text{for every (finite)} \ \Pi \subseteq S.$$

Let $\alpha \in \bigcap_{X \subseteq S} Cn(R_X, A \cup Sb(X) \cup \Pi)$ and r be the rule determined by $\langle \Pi, \alpha \rangle$, i.e.,

$$\langle Y, \beta \rangle \in r \quad \Leftrightarrow \quad \exists_{e: \, At \to S}(Y = h^e(\Pi) \ \wedge \ \beta = h^e(\alpha)).$$

Since $\langle R_X, A \cup Sb(X) \rangle \in \text{Inv}$, we have $\beta \in Cn(R_X, A \cup Sb(X) \cup Y)$ and hence

$$Y \subseteq Cn(R, A \cup Sb(X)) \Rightarrow \beta \in Cn(R, A \cup Sb(X)), \quad \text{for every} \ \langle Y, \beta \rangle \in r.$$

Thus, it has been shown that $r \in \text{Adm}(R, A \cup Sb(X))$ for each $X \subseteq S$ and hence, by 1.59, we get $r \in \text{Der}(R \cup \{r_*\}, A)$. But $\langle R, A \rangle$ is saturated and r is structural, so $r \in \text{Der}(R, A)$ and hence $\alpha \in Cn(R, A \cup \Pi)$. $\qquad\square$

Both above theorems can be written down in the consequence formalism:

Lemma 3.92. *For each family* $\{Cn_t : t \in T\}$ *of structural and saturated consequence operations, we have* $\prod_{t \in T} Cn_t \in \text{STD}$.

Theorem 3.93 (∞). *Let* $Cn \in \text{STRUCT}$. *Then*

$$Cn \in \text{STD} \quad \Leftrightarrow \quad Cn = \prod \{Cn_1 \in \text{STRUCT} : Cn \leqslant Cn_1 \in \text{SCPL}\}.$$

One can prove, similarly as Theorem 3.13, that every consistent propositional logic can be consistently extended to a saturated one, i.e.,

Theorem 3.94. *For every* $\langle R, A \rangle \in \text{Cns}$, *there is* $\langle R_1, A_1 \rangle \in \text{Std}$ *such that* $\langle R, A \rangle \preccurlyeq \langle R_1, A_1 \rangle \in \text{Cns}$.

Proof. Let us assume that $R_1 = R \cup \mathrm{Der}(R \cup \{r_*\}, A) \cap \mathrm{Struct}$ and notice that $R_1 \cup \{r_*\} \subseteq \mathrm{Der}(R \cup \{r_*\}, A)$. Therefore, we have $\mathrm{Der}(R_1 \cup \{r_*\}, A) \cap \mathrm{Struct} \subseteq$ $\subseteq \mathrm{Der}(R \cup \{r_*\}, A) \cap \mathrm{Struct} \subseteq R_1$ and hence $\langle R_1, A \rangle \in \mathrm{Std}$. Moreover, $R \subseteq R_1$ which means that $\langle R, A \rangle \preccurlyeq \langle R_1, A \rangle$. $\qquad \square$

The saturated extension of $\langle R, A \rangle$, that is the system $\langle R_1, A_1 \rangle$, is defined above in a such way that we also establish its properties:

1. $\langle R \cup \{r_*\}, A \rangle \approx \langle R_1 \cup \{r_*\}, A_1 \rangle$,

2. $\langle R, A \rangle \in \mathrm{Inv} \quad \Leftrightarrow \quad \langle R_1, A_1 \rangle \in \mathrm{Inv}$,

3. $\langle R, A \rangle \in \mathrm{Std} \quad \Leftrightarrow \quad \langle R_1, A_1 \rangle \approx \langle R, A \rangle$,

assuming that $\langle R, A \rangle \in \mathrm{Inv}$ (or at least $r_* \in \mathrm{Adm}(R, A)$) we get also

4. $\langle R_1, A_1 \rangle \in \mathrm{SCpl} \quad \Leftrightarrow \quad \langle R \cup \{r_*\}, A \rangle \in \mathrm{SCpl}$.

The property 4. allows us to construct many saturated logics which are structurally incomplete. Namely, it suffices to find such a logic $\langle R, A \rangle \in \mathrm{Inv}$ that $\langle R \cup \{r_*\}, A \rangle \notin \mathrm{SCpl}$, then the extension (such as in the proof of Theorem 3.94) of $\langle R, A \rangle$ is saturated and is structurally incomplete. Lemma 3.92 give us another method of constructing saturated logics which are not structurally complete. Let us notice that from the inequality $\mathrm{SCpl} \neq \mathrm{Std}$ it follows that, for structural completeness, the counterpart of 3.92 does not hold.

Now we would like to look more carefully at the structure of the notion of Γ-completeness. In the forthcoming considerations we will confine ourselves to propositional systems formalized over the language $\mathscr{S}_2 = \langle S_2, \rightarrow, +, \cdot, \sim \rangle$. This assumption, however, will not be essential for most of the further results.

Lemma 3.95. $\Gamma_1 \subseteq \Gamma_2 \quad \Rightarrow \quad \Gamma_1\text{-Cpl} \subseteq \Gamma_2\text{-Cpl}$.

The easy proof is omitted. Note that such monotonicity is not true for Γ-maximality. By Definition 3.1, we obtain $\mathrm{Cpl} \subseteq \Gamma\text{-Cpl} \subseteq \mathrm{SCpl}$ for every $\Gamma \subseteq S$. Let us prove that there exist sets $\Gamma \subseteq S$ such that $\mathrm{Cpl} \neq \Gamma\text{-Cpl} \neq \mathrm{SCpl}$:

Example. Take $\Gamma = Z_2 = Cn(R_{0*}, A_2)$. We have $\langle R_0, Sb(A_2) \rangle \notin \Gamma\text{-Cpl}$, as the rule

$$r_1 = \{\langle \alpha, \beta \rangle : \alpha, \beta \in At \ \lor \ \alpha, \beta \in Z_2\}$$

is Γ-structural, and $r_1 \in \mathrm{Adm}(R_0, Sb(A_2))$, whereas $q \notin Cn(R_0, Sb(A_2) \cup \{p\})$ if p, q are different variables. Hence $r \notin \mathrm{Der}(R_0, Sb(A_2))$. On the other hand, if $R_1 = \mathrm{Adm}(R_0, Sb(A_2)) \cap \mathrm{Struct}(Z_2)$, then $\langle R_1, Sb(A_2) \rangle \in \Gamma\text{-Cpl}$ but $\sim (p \rightarrow p) \notin Cn(R_1, Sb(A_2) \cup \{p\})$ because $\sim (\alpha \rightarrow \alpha) \notin Z_2 = Cn(R_1, Sb(A_2) \cup \{\alpha\})$ for every $\alpha \in Z_2$. Thus, we get $\langle R_1, Sb(A_2) \rangle \notin \mathrm{Cpl}$.

Lemma 3.96.

(i) If $e_1, \ldots, e_n \colon At \rightarrow \Gamma$ and $n \geqslant 1$, then

$$h^{e_n} \circ \ldots \circ h^{e_1}(\Gamma)\text{-Cpl} \subseteq \Gamma\text{-Cpl};$$

(ii)   *If $\Gamma \subseteq S \setminus At$ and $e: At \to \Gamma$, then*

$$h^e(\Gamma)\text{-Cpl} \neq \Gamma\text{-Cpl}.$$

Proof. To prove (i) it suffices to show that

$$\text{Struct}(\Gamma) \subseteq \text{Struct}\big(h^{e_n} \circ \ldots \circ h^{e_1}(\Gamma)\big)$$

for each $n \geqslant 1$, $e_1, \ldots, e_n: At \to \Gamma$. Let $r \in \text{Struct}(\Gamma)$, $e: At \to h^{e_n} \ldots h^{e_1}(\Gamma)$ and let $e_0: At \to \Gamma$ be such a mapping that $h^{e_n} \ldots h^{e_1}(e_0(\gamma)) = e(\gamma)$ for each $\gamma \in At$. Since the set S is countable, one can define this mapping effectively. As $h^{e_n} \circ \ldots \circ h^{e_1} \circ e_0 = e$ and $\langle h^{e_n} \ldots h^{e_0}(\Pi), h^{e_n} \ldots h^{e_0}(\alpha)\rangle \in r$ for every $\langle \Pi, \alpha \rangle \in r$, we get $\langle h^e(\Pi), h^e(\alpha)\rangle \in r$ which was to be proved.

(ii): Let $R = \{\Gamma\text{-}r_{Sb(h^e(\Gamma))}\}$ (see definition 3.7). We have $Cn\big(R, Sb(h^e(\Gamma))\big) = Sb\big(h^e(\Gamma)\big)$ and from 3.8 it follows that $\langle R, Sb(h^e(\Gamma))\rangle \in \Gamma\text{-Cpl}$. Moreover, it is obvious that $\langle R, Sb(h^e(\Gamma))\rangle \in \text{Inv}(\Gamma)$. We will show that $\langle R, Sb(h^e(\Gamma))\rangle \notin h^e(\Gamma)\text{-Cpl}$. Suppose that α is a formula such that $\alpha \in e(At)$ and $l(\phi) \geqslant l(\alpha)$ for each $\phi \in e(At)$ (we recall that $l(\alpha)$ is the length of the formula α). It has been assumed that $e(At) \subseteq S \setminus At$, hence $l(\alpha) < l\big(h^e(\phi)\big)$ for each $\phi \notin At$ and thus $\alpha \notin Sb\big(h^e(\Gamma)\big)$. Since $e(\gamma) = \alpha$ for some $\gamma \in At$, we conclude that $e(\gamma) \notin Cn\big(R, Sb(h^e(\Gamma))\big) = Cn\big(R, Sb(h^e(\Gamma)) \cup \{h^e(\alpha)\}\big)$. But $\langle R, Sb(h^e(\Gamma))\rangle \in \text{Inv}(\Gamma)$ and therefore $\gamma \notin Cn\big(R, Sb(h^e(\Gamma)) \cup \{\alpha\}\big)$. If we consider the rule

$$r = \{\langle \alpha, \gamma \rangle\} \cup \{\langle \phi, \psi \rangle : \phi \in S \land \psi \in Sb\big(h^e(\Gamma)\big)\}$$

then we can easily see that $r \in \text{Struct}\big(h^e(\Gamma) \cap \text{Adm}(R, Sb(h^e(\Gamma)))\big)$. On the other hand $r \notin \text{Der}\big(R, Sb(h^e(\Gamma))\big)$, hence $\langle R, Sb(h^e(\Gamma))\rangle \notin h^e(\Gamma)\text{-Cpl}$.   □

The reason for introducing new concepts of completeness for propositional logics is that Post-completeness is a strong property, too strong for the invariant classical propositional logic. If we decide to accept the notion of Γ-Cpl for some fixed $\Gamma \subseteq S$, it will be quite natural to render that $\langle R_0, Sb(A_2)\rangle \in \Gamma\text{-Cpl}$. Thereby, the case $\Gamma = Z_2$ (see the example on page 125) should be excluded from our considerations and we ought to look for another notion of completeness in the family $\{\Gamma\text{-Cpl} : \Gamma \subseteq S\}$. We know that $\langle R_0, Sb(A_2)\rangle \in S\text{-Cpl}$. Farther on, from 3.96 (ii) we get $h^e(S \setminus At)\text{-Cpl} \neq (S \setminus At)\text{-Cpl}$ and, by 3.95, we also have $(S \setminus At)\text{-Cpl} \subseteq S\text{-Cpl}$. Since, for some $e: At \to S$ and some $\alpha, \sim \beta \in Z_2$, we have $\{\alpha, \beta\}\text{-Cpl} \subseteq h^e(S \setminus At)\text{-Cpl}$ (see 3.95), we thus succeed in finding a set Γ such that $\langle R, Sb(A_2)\rangle \in \Gamma\text{-Cpl} \subsetneq S\text{Cpl}$. This is characteristic as we can prove:

Lemma 3.97.

(i) *If $\alpha, \sim \beta \in Z_2$, then $\langle R_0, Sb(A_2)\rangle \in \{\alpha, \beta\}\text{-Cpl}$;*

(ii) *If $\langle R_0, Sb(A_2)\rangle \in \Gamma\text{-Cpl}$, then there are $\alpha, \beta \in S_2$ such that $\alpha, \sim \beta \in Z_2$ and $\{\alpha, \beta\}\text{-Cpl} \subseteq \Gamma\text{-Cpl}$.*

Proof. The proof of (i) is quite similar to that of Theorem 3.71.

(ii): Note that $p \notin Cn(R_0, Sb(A_2) \cup \{q, \sim s\})$. Then $\langle \{q, \sim s\}, p \rangle \notin \Gamma\text{-}r_{Z_2}$ by 3.8. Hence there are $e_1, \ldots, e_n \colon At \to \Gamma$ such that for $\alpha = h^{e_n} \ldots h^{e_1}(q)$ and $\beta = h^{e_n} \ldots h^{e_1}(s)$ we have $\alpha, \sim \beta \in Z_2$. The inclusion $\{\alpha, \beta\}\text{-Cpl} \subseteq \Gamma\text{-Cpl}$ follows now from Lemma 3.96. $\qquad\square$

Thus, we know that there are $\Gamma \subseteq S_2$ such that

$$\text{Cpl} \subsetneq \Gamma\text{-Cpl} \subsetneq \text{SCpl} \quad \text{and} \quad \langle R_0, Sb(A_2) \rangle \in \Gamma\text{-Cpl}.$$

In connection with the above lemma one can expect that there exists the smallest (with respect to the relation of inclusion) or at least a minimal completeness for $\langle R_0, Sb(A_2) \rangle$. Unfortunately, this conjecture is false:

Theorem 3.98. *The family* $\{\Gamma\text{-Cpl} : \langle R_0, Sb(A_2) \in \Gamma\text{-Cpl}\}$ *does not contain any minimal element.*

Proof. Let $\langle R_0, Sb(A_2) \in \Gamma\text{-Cpl}$. By 3.97, $\langle R_0, Sb(A_2) \in \{\alpha, \beta\}\text{-Cpl} \subseteq \Gamma\text{-Cpl}$ for some $\alpha, \beta \in S_2$ such that $\alpha, \sim \beta \in Z_2$. Let $e \colon At \to \{\alpha, \beta\}$. Then we get $h^e(\alpha), \sim h^e(\beta) \in Z_2$ and hence $\langle R_0, Sb(A_2) \in h^e(\{\alpha, \beta\})\text{-Cpl}$. Moreover, by 3.96 (ii), we have $h^e(\{\alpha, \beta\})\text{-Cpl} \subsetneq \{\alpha, \beta\}\text{-Cpl} \subseteq \Gamma\text{-Cpl}$. $\qquad\square$

Despite this negative result it is still possible that there exists a minimal element in the family $\{\Gamma\text{-Cpl} \cap \text{Inv} : \langle r_0, Sb(A_2) \in \Gamma\text{-Cpl}\}$.

For any $\langle R, A \rangle$, one can define a binary relation on S_2 by

$$\alpha \sim_{RA} \beta \quad \text{iff} \quad (\alpha \equiv \beta) \in Cn(R, A).$$

We write $\sim$ instead of $\sim_{RA}$ if the system $\langle R, A \rangle$ is fixed. Let us recall that $\langle R, A \rangle$ is a system with equivalence ($\langle R, A \rangle \in \text{Equiv}$) iff $\sim$ is a congruence on the Lindenbaum matrix $\mathfrak{M}^{R, A \cup X}$ for each $X \subseteq S_2$.

Lemma 3.99. *Let* $\langle R, A \rangle \in \text{Inv} \cap \text{Equiv}$ *and* $\langle R, A \rangle \in \{a_0, \ldots, a_k\}\text{-Cpl}$ *for some* $\alpha_0, \ldots, \alpha_k \in S_2$ *(where $k \geqslant 0$). Then there are, at most, $k + 1$-element matrices* $\mathfrak{M}_1, \ldots, \mathfrak{M}_s$ *($s \geqslant 1$) such that $\mathfrak{M}_1 \times \ldots \times \mathfrak{M}_s$ is strongly adequate for $\langle R, A \rangle$.*

Proof. Let us accept the assumptions of our lemma. Then, of course, $\langle R, A \rangle$ is structurally complete (see Lemma 3.95) and the relation $\sim_{RA}$ is a congruence relation on the Lindenbaum matrix $\mathfrak{M}^{R, A}$. Thus,

1. $Cn(R, A) = E(\mathfrak{M}^{R, A}) = E(\mathfrak{M}^{R, A} / \sim)$.

We can assume $Cn(R, A) \neq S_2$, as for the inconsistent logic a strongly adequate matrix can be easily found. The main step in our proof is to show that

(2) $Cn(R, A) = \bigcap \{E(\mathfrak{M}) : \mathfrak{M} \subseteq \mathfrak{M}^{R, A} / \sim \ \wedge \ Nc(\mathfrak{M}) \leqslant k + 1\}$.

The inclusion ($\subseteq$) is obvious. We have to prove the reverse inclusion. We do not know if there ever exists a matrix $\mathfrak{M}$ which fulfills the above conditions; but it is so indeed. Suppose that $\phi \notin Cn(R, A)$ and let

(3) $A^0 = At(\phi)$,
$\quad A^{i+1} = A^i \cup \{\sim \alpha : \alpha \in A^i\} \cup \{\alpha_1 \to \alpha_2 : \alpha_1, \alpha_2 \in A^i\} \cup \{\alpha_1 + \alpha_2 : \alpha_1, \alpha_2 \in A^i\} \cup \{\alpha_1 \cdot \alpha_2 : \alpha_1, \alpha_2 \in A^i\}$.

Let $g : A^{k+1} \to At$ be any one-to-one mapping such that $g(\gamma) = \gamma$ for each $\gamma \in At(\phi)$. In the sequel we will write p_α instead of $g(\alpha)$ for $\alpha \in A^{k+1}$. We have $p_\gamma = \gamma$ if $\gamma \in At(\phi)$. Moreover, let

(4) $\Pi_i = \{\alpha \equiv p_\alpha : \alpha \in A^i\}$ for $i = 0, \ldots, k+1$.

Obviously, we have $\Pi_0 \subseteq \Pi_1 \subseteq \ldots \subseteq \Pi_{k+1}$ and all these sets are finite. What is more $At(\phi) \subseteq At(\Pi_t) = \{p_\alpha : \alpha \in A^i\}$. Take $e : At \to S$ such that $e(p_\alpha) = \alpha$ for each $\alpha \in A^{k+1}$. Then $e(\gamma) = \gamma$ for each $\gamma \in At(\phi)$ and hence $h^e \phi = \phi$. Thus, we get $h^e(\phi) \notin Cn(R, A)$ and $h^e(\Pi_{k+1}) \subseteq Sb(p \equiv p) \subseteq Cn(R, A)$ and hence $\phi \notin Cn(R, A \cup \Pi_{k+1})$ as $\langle R, A \rangle \in$ Inv. Then Γ-completeness of $\langle R, A \rangle$, where $\Gamma = \{\alpha_0, \ldots, \alpha_k\}$, yields by 3.8 $\langle \Pi_{k+1}, \phi \rangle \notin \Gamma\text{-}r_{Cn(R,A)}$. Suppose that $\Pi_{k+1} \subseteq Cn(R, A)$. Then $Cn(R, A) \neq \emptyset$ and $p_{\sim\gamma} \equiv \sim \gamma \in \Pi_1 \subseteq Cn(R, A)$ for each $\gamma \in At(\phi)$. Since $p_\gamma \in At$ and $p_{\sim\gamma} \neq \gamma$ we get $\alpha \equiv \sim \gamma \in Cn(R, A)$ for each $\alpha \in Cn(R, A)$. But $\langle R, A \rangle \in$ Equiv, thus $\sim \gamma \in Cn(R, A)$ and this implies $p_{\sim\gamma} \in Cn(R, A)$ (by the fact that $p_{\sim\gamma} \equiv \sim \gamma \in Cn(R, A)$). Therefore, some variable belongs to $Cn(R, A)$ and hence $Cn(R, A) = S_2$, which contradicts our assumptions. Thus we have proved that $\langle \Pi_{k+1}, \phi \rangle \notin \Gamma\text{-}r_{Cn(R,A)}$ and $\Pi_{k+1} \nsubseteq Cn(R, A)$. By 3.7, there are $e_1, \ldots, e_n : At \to \Gamma$ (where $n \geqslant 1$) such that

(5) $h^{e_n} \ldots h^{e_1}(\Pi_{k+1}) \subseteq Cn(R, A)$ and $h^{e_n} \ldots h^{e_1}(\phi) \notin Cn(R, A)$.

Since $\Gamma = \{\alpha_0, \ldots, \alpha_k\}$, $Nc(h^{e_n} \ldots h^{e_1}(At)) \leqslant k+1$ and hence in the sequence

$$h^{e_n} \ldots h^{e_1}(At(\Pi_0)) \subseteq h^{e_n} \ldots h^{e_1}(At(\Pi_1)) \subseteq \ldots \subseteq h^{e_n} \ldots h^{e_1}(At(\Pi_{k+1}))$$

there are two equal elements, i.e.,

(6) $h^{e_n} \ldots h^{e_1}(At(\Pi_i)) = h^{e_n} \ldots h^{e_1}(At(\Pi_{i+1}))$ for some $i \leqslant k$.

Take $B = \{[\beta]_\sim : h^{e_n} \ldots h^{e_1}(At(\Pi_i))\}$ and prove that B expands a submatrix of $\mathfrak{M}^{R,A}/\sim$. Let $[\beta_1], [\beta_2] \in B$, i.e., let $\beta_1 = h^{e_n} \ldots h^{e_1}(p_{\alpha_1})$ and $\beta_2 = h^{e_n} \ldots h^{e_1}(p_{\alpha_2})$ for some $\alpha_1, \alpha_2 \in A_i$. Then $\alpha_1 \to \alpha_2 \in A_{i+1}$ and hence $p_{\alpha_1 \to \alpha_2} \equiv (\alpha_1 \to \alpha_2) \in \Pi_{i+1} \subseteq \Pi_{k+1}$. By (5), we get

$$h^{e_n} \ldots h^{e_1}(p_{\alpha_1 \to \alpha_2}) \equiv (\beta_1 \to \beta_2) \in Cn(R, A)$$

and thus, by (6) and (3),

$$[\beta_1 \to \beta_2] \in B.$$

Similarly, it can be shown that $[\beta_1 + \beta_2], [\beta_1 \cdot \beta_2], [\sim \beta_1] \in B$. Therefore B is closed under the operations from the matrix $\mathfrak{M}^{R,A}/\sim$ and then it expands a submatrix $\mathfrak{M} = \langle \mathscr{B}, B^* \rangle$ of $\mathfrak{M}^{R,A}/\sim$. Moreover, the set B contains at most $k+1$ elements as $Nc(h^{e_n} \ldots h^{e_1}(At(\Pi_i))) \leqslant k+1$. Consider the mapping $v : At \to B$ such that $v(\gamma) = [h^{e_n} \ldots h^{e_1}(\gamma)]$ for each $\gamma \in At(\phi)$ (note that $At(\phi) \subseteq At(\Pi_i)$).

Obviously, $h^v(\phi) = [h^{e_n} \ldots h^{e_1}(\phi)]$, where $h^v \colon \mathscr{S}_2 \to \mathscr{B}$, and hence (by (5) and (1)) $h^v(\phi) \notin B^*$. Thus, $\phi \notin E(\mathfrak{M})$ which completes the proof of (2).

There are only finitely many non-isomorphic matrices with at most $k+1$ elements. Therefore, we can find $\mathfrak{M}_1, \ldots, \mathfrak{M}_s$ (where $s \geqslant 1$) at most $k+1$-element submatrices of $\mathfrak{M}^{R,A}/\!\sim$ such that

(7) $Cn(R, A) = E(\mathfrak{M}_1) \cap \ldots \cap E(\mathfrak{M}_s)$.

We know that $\sim$ is a congruence-relation in the Lindenbaum matrix $\mathfrak{M}^{R,A}$. Thus $\overrightarrow{\mathfrak{M}^{R,A}} = \overrightarrow{\mathfrak{M}^{R,A}/\!\sim} \leqslant \overrightarrow{\mathfrak{M}_i}$ by 2.37 and 2.39, and $Cn_{RA} \leqslant \overrightarrow{\mathfrak{M}^{R,A}}$. Hence, by 2.41, we get $Cn(R, A \cup X) \subseteq \mathbf{P}_{i \leqslant s} \overrightarrow{\mathfrak{M}_i}(X)$ for each $X \subseteq S_2$. On the other hand, we have $\mathbf{P}_{i \leqslant s} \overrightarrow{\mathfrak{M}_i}(X) \subseteq Cn(R, A \cup X)$ by 3.59 and (7) (Let us note that $\mathbf{P}_{i \leqslant s} \overrightarrow{\mathfrak{M}_i}$ is finite and hence the corresponding matrix consequence is finitistic). Thus, we conclude that the product $\mathfrak{M}_1 \times \ldots \times \mathfrak{M}_s$ is strongly adequate for $\langle R, A \rangle$. $\square$

Theorem 3.100. *Let $\langle R, A \rangle \in \mathrm{Inv} \cap \mathrm{Equiv}$. Then $\langle R, A \rangle \in \Gamma\text{-Cpl}$ for a finite set $\Gamma \subseteq S$ iff $\langle R, A \rangle \in \mathrm{SCpl}$ and there is a finite adequate matrix $\mathfrak{M}$ for $\langle R, A \rangle$.*

Proof. The implication ($\Rightarrow$) follows from the above lemma. To prove the reverse implication ($\Leftarrow$) let us assume that $\mathfrak{M} = \langle \mathscr{B}, B^* \rangle$ is a finite matrix adequate for $\langle R, A \rangle \in \mathrm{SCpl}$ and let $\phi \notin Cn(R, A \cup X)$ for some (finite) set $X \subseteq S_2$. Then, by 3.64, we have $\langle X, \phi \rangle \notin r_{Cn(R,A)}$ and hence there exists a substitution $e \colon At \to S_2$ such that $h^e(X) \subseteq Cn(R, A)$ and $h^e(\phi) \notin Cn(R, A)$. Suppose that $B = \{a_0, \ldots, a_k\}$ (B is the base of $\mathfrak{M}$). Since $\mathfrak{M}$ is adequate for $\langle R, A \rangle$, we have $h^v(h^e(\phi)) \notin B^*$ for some $v \colon At \to B$. Define $e_1 \colon At \to \{p_0, \ldots, p_k\}$ by

$$e_1(\gamma) = p_i \quad \text{iff} \quad v(\gamma) = a_i$$

and take $v_1 \colon At \to B$ such that $v_1(p_i) = a_i$ for each $i = 0, \ldots, k$. We have $v_1(e_1(\gamma)) = v(\gamma)$ for each $\gamma \in At$ and hence $h^{v_1}(h^{e_1}(h^e(\phi))) \notin B^*$. Thus, we get $h^{e_1}(h^e(\phi)) \notin Cn(R, A)$ and $(h^{e_1}(h^e(X)) \subseteq Cn(R, A)$. It has been assumed that $\sim_{RA}$ is a congruence on the Lindenbaum matrix $\mathfrak{M}^{R,A}$. Observe that this relation divides the set of all formulas built up from the variable $p_0, \ldots, p_k$ into a finite number of equivalence classes. Indeed, each formula α with $At(\alpha) \subseteq \{p_0, \ldots, p_k\}$ determines a mapping $f_\alpha \colon B^{k+1} \to B$ by $f_\alpha(v(p_0), \ldots, v(p_k)) = h^v(\alpha)$ for every $v \colon At \to B$. Since B is finite, there exists a finite set Γ of formulas such that $f_\alpha \in \{f_\gamma : \gamma \in \Gamma\}$ for each α. Thus, for each formula α built up from variables $p_0, \ldots, p_k$ (i.e., $At(\alpha) \subseteq \{p_0, \ldots, p_k\}$) there is $\gamma \in \Gamma$ such that $h^v(\alpha) = h^v(\gamma)$ for each $v \colon At \to B$. But $p \equiv p \in Cn(R, A)$ and hence $\alpha \equiv \gamma \in Cn(R, A)$. Thus, Γ contains representatives of all equivalence classes into which the subalgebra generated by $p_0, \ldots, p_k$ is divided. Since $e_1 \colon At \to \{p_0, \ldots, p_k\}$, we can find a mapping $e_2 \colon At \to \Gamma$ such that $e_2(\gamma) \equiv h^{e_1}(e(\gamma)) \in Cn(R, A)$ for every $\gamma \in At$. Then $h^{e_2}(\phi) \sim h^{e_1}(h^e(\phi))$ and hence (recall that $h^{e_1}h^e(\phi) \notin Cn(R, A)$), $h^{e_2}(\phi) \notin Cn(R, A)$. By the same argument we get $h^{e_2}(X) \subseteq Cn(R, A)$. It has been shown that $\langle X, \phi \rangle \notin \Gamma\text{-}r_{Cn(R,A)}$ for any X, ϕ such that $\phi \notin Cn(R, A \cup X)$. This means that $\Gamma\text{-}r_{Cn(R,A)} \in \text{Der}(R, A)$ and hence $\langle R, A \rangle \in \Gamma\text{-Cpl}$. $\qquad \square$

Corollary 3.101. *Let $\langle R, A \rangle \in \text{Inv} \cap \text{Equiv}$. Then $\langle R, A \rangle \in \Gamma\text{-Cpl}$ for a finite set $\Gamma \subseteq S$ iff $\langle R, A \rangle \in \text{SCpl}$ and there exists a finite matrix $\mathfrak{M}$ such that $Cn_{RA} = \overrightarrow{\mathfrak{M}}$.*

All the above results expose certain connections between adequacy (weak and strong) and the notion of a finite Γ-completeness. Moreover, by Theorem 3.98, we can give a partial positive answer to the question on the existence of a minimal Γ-completeness for $\langle R_0, Sb(A_2) \rangle$ in the family of invariant calculi with equivalence. Note that the family $\Gamma\text{-Cpl} \cap \text{Inv} \cap \text{Equiv}$, for a finite set Γ, contains only a finite number of non-equivalent systems. Therefore, we can choose such $\alpha, \beta \in S_2$ that $\alpha, \sim \beta \in Z_2$ (hence $\langle R_0, Sb(A_2) \rangle \in \{\alpha, \beta\}\text{-Cpl}$) and that

$$\Gamma\text{-Cpl} \cap \text{Inv} \cap \text{Equiv} \subsetneq \{\alpha, \beta\}\text{-Cpl} \cap \text{Inv} \cap \text{Equiv} \ \Rightarrow \ \langle R_0, Sb(A_2) \rangle \notin \Gamma\text{-Cpl}$$

for every $\Gamma \subseteq S$. One can even find such a set $\Gamma \subseteq S$ that

$$\langle R, A \rangle \in \Gamma\text{-Cpl} \cap \text{Inv} \cap \text{Equiv} \quad \text{iff} \quad \langle R, A \rangle \approx \langle R_0, Sb(A_2) \rangle.$$

We can also give a characterization of a two-valued logic without any characterization of the propositional connectives. We say that $\langle R, A \rangle$ is *two-valued* iff there exists a two-element matrix $\mathfrak{M} = \langle \mathcal{B}, \{1\} \rangle$, where $B = \{0, 1\}$, such that $Cn(R, A) = E(\mathfrak{M})$ and $N(\mathfrak{M}) \subseteq \text{Der}(R, A)$. Let S_p be the sublanguage of S_2 generated by the variable p, i.e., $\alpha \in S_p$ iff $At(\alpha) = \{p\}$. We have

Theorem 3.102. *Suppose that $\langle R, A \rangle \in \text{Equiv}$. Then $\langle R, A \rangle$ is two-valued iff $\langle R, A \rangle \in \{\alpha, \beta\}\text{-Cpl}$ for some $\alpha, \beta \in S$ and one of the following conditions holds:*

(i) $S_p/\sim_{RA} = \{[p], [p \to p]\}$,

(ii) $\alpha(p \equiv p) \notin Cn(R, A)$ *for some $\alpha \in S_p$.*

The proof of this theorem is similar to that of Theorem 3.99. Thus, two-valued propositional logics can be characterized by means of $\{\alpha, \beta\}$-completeness. The conditions (i),(ii) are rather technical. They state that the negation is definable in $\langle R, A \rangle$ (condition (ii)) or that all formulas in one variable are equivalent either to p or to $p \equiv p$ (condition (i)).

Chapter 4

Characterizations of propositional connectives

Our attempt is to define propositional logics by use of certain conditions, the so-called Cn-definitions, which characterize basic properties of connectives involved in these logics. Our approach turns out to be successful in the case of intuitionistic logic but not quite satisfactory for the classical logic. In our opinion, there is still the need for a complete and adequate set of postulates which would characterize basic properties of classical connectives.

4.1 Cn-definitions

A characterization of the classical logic in terms of a consequence operation is due to Tarski [117] and [118], 1930. His ideas have been elaborated by Grzegorczyk [37], 1972, and Pogorzelski, Słupecki [87], 1960. So, it is well known that Cn_2 is the least consequence operation Cn which fulfills the following conditions, see Corollary 1.67, for every $X, Y \subseteq S_2$ and every $\alpha, \beta \in S_2$:

$$
\begin{array}{lll}
\text{(T0)} & X \subseteq Cn(Y) \;\Rightarrow\; Y \cup Cn(X) \subseteq Cn(Y); \\
\text{(T1)} & Cn(\{\alpha, \sim \alpha\}) = S_2 \;\;\&\;\; Cn(\{\alpha\}) \cap Cn(\{\sim \alpha\}) = Cn(\emptyset); \\
\text{(T2)} & \alpha \to \beta \in Cn(X) \;\Leftrightarrow\; \beta \in Cn(X \cup \{\alpha\}); \\
\text{(T3)} & Cn(X \cup \{\alpha + \beta\}) = Cn(X \cup \{\alpha\}) \cap Cn(X \cup \{\beta\}); \\
\text{(T4)} & Cn(X \cup \{\alpha \cdot \beta\}) = Cn(X \cup \{\alpha, \beta\}).
\end{array}
$$

It is also known that Cn_2 is the only consistent and structural consequence operation satisfying the above conditions. The above characterization of the classical logic is not quite satisfactory. Firstly the conditions are not quite uniform. The condition (T0) postulates that Cn is a consequence operation. We would prefer to have, in addition to (T0), one condition of the form :

$$
\alpha \bigtriangleup \beta \in Cn(X) \qquad \Leftrightarrow \qquad \cdots
$$

for each $\triangle \in \{\sim, \to, +, \cdot\}$. The right-hand side of each equivalence should satisfy the usual definition conditions. It means that the connective occurring on the left-hand side should not appear on the right-hand side of the corresponding equivalence. There should not be *circulus vitiosus* in definitions (for instance, it should not be allowed that $\to$ is defined by means of conditions involving $\sim$ and, vice versa, $\sim$ is defined by use of $\to$). The conditions should be as uniform as it is possible. Equivalences of this form will be called *Cn-definitions of propositional connectives*. We do not specify (meta)logical means allowed there (on the right-hand sides) but, instead, we illustrate our approach with an example:

(H1) $\sim \alpha \in Cn(X) \quad \Leftrightarrow \quad S_2 \subseteq Cn(X \cup \{\alpha\});$

(H2) $\alpha \to \beta \in Cn(X) \quad \Leftrightarrow \quad Cn(X \cup \{\beta\}) \subseteq Cn(X \cup \{\alpha\});$

(H3) $\alpha + \beta \in Cn(X) \quad \Leftrightarrow \quad Cn(X \cup \{\alpha\}) \cap Cn(X \cup \{\beta\}) \subseteq Cn(X);$

(H4) $\alpha \cdot \beta \in Cn(X) \quad \Leftrightarrow \quad Cn(X \cup \{\alpha\}) \cup Cn(X \cup \{\beta\}) \subseteq Cn(X).$

As is well known, see Corollary 1.66 (and following comments), we have:

Theorem 4.1. *The intuitionistic consequence operation Cn_i is the least one which fulfills the above conditions* (H).

Thus, the intuitionistic logic possess a satisfactory and complete characterization by use of specific properties of propositional connectives. Our task is to find an appropriate set of Cn-definitions for the classical logic Cn_2. And, in principle, we would like to achieve this goal without changing the general properties of consequence operations given by the condition (T0).

We recall that Cn_2 is the consequence operation determined by the (invariant version of) the classical logic $\langle R_0, Sb(A_2) \rangle$. We have $R_0 = \{r_0\}$, where r_0 is the modus ponens rule, and $Sb(A_2)$ consists of the following axiom schemata:

(1) $\alpha \to (\beta \to \alpha)$

(2) $[\alpha \to (\alpha \to \beta)] \to (\alpha \to \beta)$

(3) $(\alpha \to \beta) \to [(\beta \to \gamma) \to (\alpha \to \gamma)]$

(4) $\alpha \to \alpha + \beta$

(5) $\beta \to \alpha + \beta$

(6) $(\alpha \to \gamma) \to [(\beta \to \gamma) \to (\alpha + \beta \to \gamma)]$

(7) $\alpha \cdot \beta \to \alpha$

(8) $\alpha \cdot \beta \to \beta$

(9) $(\alpha \to \beta) \to [(\alpha \to \gamma) \to (\alpha \to \beta \cdot \gamma)]$

(10) $\alpha \to (\sim \alpha \to \beta)$

(11) $(\alpha \to \sim \alpha) \to \sim \alpha$

(12) $\sim\sim \alpha \to \alpha$

for each $\alpha, \beta, \gamma \in S_2$. Let $Z_2 = Cn_2(\emptyset)$ be the set of (classical) tautologies over S_2.

4.2 The system (D)

Firstly, let us consider the following set of Cn-definitions called the system (D):

$$\text{(D1)} \quad \sim \alpha \in Cn(X) \quad \Leftrightarrow \quad Cn(X \cup \{\alpha\}) = S_2;$$
$$\text{(D2)} \quad \alpha \to \beta \in Cn(X) \quad \Leftrightarrow \quad Cn(X \cup \{\alpha, \sim \beta\}) = S_2;$$
$$\text{(D3)} \quad \alpha + \beta \in Cn(X) \quad \Leftrightarrow \quad Cn(X \cup \{\sim \alpha, \sim \beta\}) = S_2;$$
$$\text{(D4)} \quad \alpha \cdot \beta \in Cn(X) \quad \Leftrightarrow \quad Cn(X \cup \{\sim \alpha\}) \cap Cn(X \cup \{\sim \beta\}) = S_2.$$

The operation Cn_2 clearly satisfies the conditions (D). Note that, if a consequence operation Cn satisfies these conditions, then so does any of its axiomatic extensions. Thus, the conditions do not define a consequence operation in a unique way. It also means that any Cn satisfying the conditions does not need to be finitary, nor structural. If one takes, however, the least (D)-operation (perceived as the intersection of all operations satisfying these conditions), then this operation — defined in a unique way — is finitary and structural. So, let us try to identify the least consequence operation satisfying the above conditions.

Lemma 4.2. *If a consequence operation Cn fulfills* (D1)–(D4), *then*

(i) $\beta \in Cn(\{\alpha, \sim \alpha\})$, *for every* $\alpha, \beta \in S_2$;

(ii) $\beta \in Cn(\{\alpha, \alpha \to \beta\})$, *for every* $\alpha \in S_2$ *and* $\beta \in S_2 \setminus At$;

(iii) $\beta \in Cn(X \cup \{\alpha\}) \Rightarrow (\alpha \to \beta) \in Cn(X)$, *for every* $\alpha, \beta \in S_2$, $X \subseteq S_2$.

Proof. (i): Follows immediately from (D1) and $\sim \alpha \in Cn(\{\sim \alpha\})$.

(ii): If $\beta \notin At$, then β is $\triangle(\gamma, \delta)$ with $\triangle \in \{\sim, \to, \cdot, +\}$. As Cn fulfills the conditions (D), there are formulas β_i with $i \in I \subseteq \{1, 2\}$ such that

$$\beta \in Cn(X) \quad \Leftrightarrow \quad Cn(X \cup \{\beta_i\}_{i \in I}) = S_2$$

for every $X \subseteq S_2$. Thus, $Cn(\{\beta\} \cup \{\beta_i\}_{i \in I}) = S_2$ which shows $\sim \beta \in Cn(\{\beta_i\}_{i \in I})$ by (D1). Since we have $Cn(\{\alpha \to \beta, \alpha, \sim \beta\}) = S_2$ by the condition (D2), we also get $Cn(\{\alpha \to \beta, \alpha\} \cup \{\beta_i\}_{i \in I}) = S_2$ and hence $\beta \in Cn(\{\alpha \to \beta, \alpha\})$.

(iii): Suppose that $\beta \in Cn(X \cup \{\alpha\})$. Then $Cn(X \cup \{\alpha, \sim \beta\}) = S_2$ by (i) and hence $(\alpha \to \beta) \in Cn(X)$ by (D2). $\qquad \square$

Lemma 4.3. *If a consequence operation Cn over the language S_2 fulfills the conditions* (D), *then* $Sb(A_2) \subseteq Cn(\emptyset)$.

Proof. (1): Easily follows from Lemma 4.2 (iii).

(2): Since $(\alpha \to \beta) \in Cn(\{\alpha, \alpha \to (\alpha \to \beta)\})$ by Lemma 4.3 (ii), we get $Cn(\{\alpha, \sim \beta, \alpha \to (\alpha \to \beta)\}) = S_2$ by (D2). Hence $(\alpha \to \beta) \in Cn(\{\alpha \to (\alpha \to \beta)\})$ and (Ax2) follows from Lemma 4.2 (iii).

(3): We have $Cn(\{\beta \to \gamma, \beta, \sim \gamma\}) = S_2$ by (D2) and hence we also get $\sim \beta \in Cn(\{\beta \to \gamma, \sim \gamma\})$ by (D1). Since $Cn(\{\alpha \to \beta, \alpha, \sim \beta\}) = S_2$ by (D2), we get $Cn(\{\alpha \to \beta, \beta \to \gamma, \alpha, \sim \gamma\}) = S_2$ and hence $(\alpha \to \gamma) \in Cn(\{\alpha \to \beta, \beta \to \gamma\})$ on the basis of (D2).

(4) and (5): Follow from (D3) and Lemma 4.2.

(6): Since $Cn(\{\alpha \to \gamma, \alpha, \sim \gamma\}) = S_2 = Cn(\{\beta \to \gamma, \beta, \sim \gamma\})$, we also get $\sim \alpha \in Cn(\{\alpha \to \gamma, \sim \gamma\})$ and $\sim \beta \in Cn(\{\beta \to \gamma, \sim \gamma\})$ by (D1). Hence

$$Cn(\{\alpha \to \gamma, \beta \to \gamma, \sim \gamma, \alpha + \beta\}) = S_2$$

by (D3) which shows that $(\alpha + \beta \to \gamma) \in Cn(\{\alpha \to \gamma, \beta \to \gamma\})$.

(7) and (8): Since $Cn(\{\alpha \cdot \beta, \sim \alpha\}) = S_2 = Cn(\{\alpha \cdot \beta, \sim \beta\})$ by (D4), we get (7) and (8) immediately by (D2).

(9): We get $Cn(\{\alpha \to \beta, \alpha, \sim \beta\}) = Cn(\{\alpha \to \gamma, \alpha, \sim \gamma\}) = S_2$ by (D2). Hence $\beta \cdot \gamma \in Cn(\{\alpha \to \beta, \alpha \to \gamma, \alpha\})$ by (D4). Then, to get (9) it suffices to use Lemma 4.2 (iii).

(10): Follows from Lemma 4.2 (i) and (iii).

(11): We get $\sim \alpha \in Cn(\{\alpha, \alpha \to \sim \alpha\})$ by Lemma 4.2 (ii). Then, by (D1), $Cn(\{\alpha, \alpha \to \sim \alpha\}) = S_2$ which gives $\sim \alpha \in Cn(\{\alpha \to \sim \alpha\}) = S_2$.

(12): Since $Cn(\{\sim \alpha, \sim\sim \alpha\}) = S_2$ we get (12) by (D2).   □

Now, the least consequence operation satisfying the conditions (D) can be identified quite easily. Let Cn_D be the consequence operation determined by the axioms $Sb(A_2)$ and the following inferential rules:

$$r_{01} : \frac{\alpha, \; \alpha \to \sim \beta}{\sim \beta}, \qquad r_{02} : \frac{\alpha, \; \alpha \to (\beta \to \gamma)}{\beta \to \gamma}, \qquad r_{03} : \frac{\alpha, \; \alpha \to \beta + \gamma}{\beta + \gamma},$$

$$r_{04} : \frac{\alpha, \; \alpha \to \beta \cdot \gamma}{\beta \cdot \gamma}, \qquad r_1 : \frac{\alpha, \; \sim \alpha}{\beta}.$$

Note that r_{0i} for $i \in \{1, 2, 3, 4\}$ are subrules of the modus ponens rule r_0. Since the rules of Cn_D are classically valid, we get $Cn_D(X) \subseteq Cn_2(X)$ for every set X. Using known properties of the classical logic (such as the deduction theorem) one easily shows that for every set $X \subseteq S_2$:

$$
\begin{array}{llll}
(*) & Cn_D(X) = S_2 & \Leftrightarrow & Cn_2(X) = S_2 \quad ; \\
(**) & \alpha \in Cn_D(X) & \Leftrightarrow & \alpha \in Cn_2(X), \quad \text{if } \alpha \notin At; \\
(***) & \alpha \in Cn_D(X) & \Leftrightarrow & \alpha \in X, \quad \text{if } \alpha \in At, \; Cn_2(X) \neq S_2.
\end{array}
$$

Since the classical logic Cn_2 fulfills the conditions (D) and those conditions involve compound formulas on the left-hand side of the equivalences, and inconsistent theories on the right-hand side, it follows from (*) and (**) that Cn_D fulfills the conditions (D), as well. Besides, by Lemma 4.2 and 4.3, we get $Cn_D \leqslant Cn$ for every consequence operation Cn fulfilling the conditions (D). Consequently

Theorem 4.4. *The operation Cn_D is the least consequence operation satisfying the conditions* (D).

To get $Cn_2 \leqslant Cn_D$, one would need to show that r_0 is a derivable rule of Cn_D. By Lemma 4.3, it would be sufficient to show that $\beta \in Cn_D(\{\alpha, \alpha \to \beta\})$

for $\beta \in At$. However, this is not true (as we shall see later on) and, consequently, Cn_2 is not the least consequence operation which fulfills (D). As no tautology is a variable, we can only get by (**):

Corollary 4.5. $Z_2 = Cn_D(\emptyset)$.

Now, let us consider a three-element matrix $\mathfrak{M}_D$, on the set $\{1, 2, 3\}$, with 3 as the only designated element, and the connectives interpreted as:

$\to$	1	2	3
1	3	3	3
2	1	3	3
3	1	3	3

$+$	1	2	3
1	1	3	3
2	3	3	3
3	3	3	3

$\cdot$	1	2	3
1	1	1	1
2	1	3	3
3	1	3	3

$\sim$	
1	3
2	1
3	1

Note that the classical two-element matrix $\mathfrak{M}_2$ is (isomorphic to) a submatrix of $\mathfrak{M}_D$. Hence, $\overrightarrow{\mathfrak{M}_D} \leqslant \overrightarrow{\mathfrak{M}_2} = Cn_2$ where $\overrightarrow{\mathfrak{M}_D}$ and $\overrightarrow{\mathfrak{M}_2}$ denote the matrix consequences (see Definition 2.32) determined by $\mathfrak{M}_D$ and $\mathfrak{M}_2$, respectively.

Notice that $q \notin \overrightarrow{\mathfrak{M}_D}(p \to q, p)$ for all distinct $p, q \in At$ (take the valuation v such that $v(p) = 3$ and $v(q) = 2$) and hence $\overrightarrow{\mathfrak{M}_D} < Cn_2$. It means, in particular, that the Cn-definitions (D) do not determine the classical connectives but connectives of a weaker logic. We also get

Theorem 4.6. $\overrightarrow{\mathfrak{M}_D} = Cn_D$.

Proof. We prove that $\overrightarrow{\mathfrak{M}_D} \leqslant Cn_D$ as the reverse is quite obvious — it suffices to notice that the axioms and rules of Cn_D are valid in $\mathfrak{M}_D$.

Let $\alpha \notin Cn_D(X)$. If $\alpha \notin At$, then $\alpha \notin \overrightarrow{\mathfrak{M}_2}(X)$ by (**) and hence we get $\alpha \notin \overrightarrow{\mathfrak{M}_D}(X)$ as $\mathfrak{M}_2$ is a submatrix of $\mathfrak{M}_D$.

Suppose that $\alpha \in At$. If $Cn_2(X) = S_2$, then $Cn_D(X) = S_2$ by (*) contradicting our assumptions. We can assume, therefore, that X is satisfiable in the classical two-element matrix. Hence $h^v(X) \subseteq \{3\}$ for a valuation $v \colon At \to \{1, 3\}$. We define a valuation $w \colon At \to \{1, 2, 3\}$ by

$$w(p) = \begin{cases} 2 & \text{if } p \text{ is } \alpha \text{ and } v(p) = 3 \\ v(p) & \text{otherwise.} \end{cases}$$

One shows by induction on the length of a formula that $h^v(\phi) = h^w(\phi)$ for every ϕ different from α. We also get $w(\alpha) = 2$. Since $\alpha \notin Cn_D(X)$, we have $\alpha \notin X$ and hence $h^w(X) = h^v(X) \subseteq \{3\}$ which shows $\alpha \notin \overrightarrow{\mathfrak{M}_D}(X)$. $\square$

Thus, $Cn_D < Cn_2$. Note that Cn_D is very close to the classical logic. In particular, $Cn_D(\emptyset) = Z_2$ and all instances of r_0 with compound conclusions are Cn_D-valid. There are only lacking (classically valid) inferences of the form X/α where α is a variable and $\alpha \notin X$, e.g., $\alpha \cdot \alpha/\alpha$. Let us prove additionally that Cn_D is a *threshold* logic which means that Cn_2 is the only structural and consistent (proper) extension of Cn_D:

Theorem 4.7. *If Cn is a structural and consistent consequence operation and $Cn > Cn_D$, then $Cn = Cn_2$.*

Proof. Let $Cn \in \text{STRUCT} \cap \text{CNS}$ and $Cn > Cn_D$. Then $\overrightarrow{\mathfrak{M}_D} < Cn$ by Theorem 4.6. Since we have $Cn(\emptyset) = \bigcap\{\overrightarrow{\mathfrak{N}}(\emptyset) : \mathfrak{N} \subseteq \mathfrak{M}_D \wedge Cn(\emptyset) \subseteq \overrightarrow{\mathfrak{N}}(\emptyset)\}$ (see Theorem 2.38), and $\mathfrak{M}_2$ is the only proper submatrix of $\mathfrak{M}_D$, we get $Cn(\emptyset) = \overrightarrow{\mathfrak{M}_D}(\emptyset) = Z_2$. As Cn_2 is structurally complete we conclude that $Cn \leqslant Cn_2$, see Corollary 3.59.

Thus $Cn_D < Cn \leqslant Cn_2$. Moreover, by the conditions $(*)$–$(***)$ above, $\alpha \in Cn_2(X)$ yields $\alpha \in Cn(X)$ if $\alpha \notin At$, or $\alpha \in X$, or $Cn_2(X) = S_2$. To get $Cn = Cn_2$ it suffices now to show that $\alpha \in Cn(X)$ for every X, α such that $\alpha \in At \setminus X$ and $\alpha \in Cn_2(X) \neq S_2$.

Let $\alpha \notin Cn(X)$ for every X, α such that $\alpha \in At \setminus X$ and $\alpha \in Cn_2(X) \neq S_2$. Since $Cn_D < Cn$, we get $\beta \in Cn(Y) \setminus Cn_D(Y)$ for some β, Y. Then, according to $(*)$ and the above assumption, we must have $\beta \notin At$. Hence, by $(**)$, we would get $\beta \notin Cn_2(X)$ contradicting our assumptions. Thus, we conclude that there are X_0, α_0 such that $\alpha_0 \in At \setminus X_0$, and $\alpha_0 \in Cn_2(X_0) \neq S_2$, and $\alpha_0 \in Cn(X_0)$.

As $Cn_2(X_0) \neq S_2$, the set X_0 is satisfiable in $\mathfrak{M}_2$ and hence $h^v(X_0) \subseteq \{3\}$ for a valuation $v\colon At \to \{1,3\}$. Let us define a substitution $e\colon At \to S_2$ by

$$e(p) = \begin{cases} \alpha_0 & \text{if } p \text{ is } \alpha_0 \\ \alpha_0 \to \alpha_0 & \text{if } v(p) = 3 \text{ and } p \text{ is not } \alpha_0 \\ \sim(\alpha_0 \to \alpha_0) & \text{if } v(p) = 1 \text{ and } p \text{ is not } \alpha_0. \end{cases}$$

Note that $Cn_2(h^e(X_0)) \neq S_2$ and no element of $h^e(X_0)$ is a variable. Besides, $\alpha_0 = h^e(\alpha_0) \in Cn_2(h^e(X_0))$. Thus, each element of $h^e(X_0)$ is either valid in classical logic or equivalent to α_0. So $h^e(X_0) \subseteq Cn_2(\{(\alpha_0 \to \alpha_0) \to \alpha_0\})$ and hence, by $(**)$ and Theorem 4.6, we get

$$h^e(X_0) \subseteq \overrightarrow{\mathfrak{M}_D}(\{(\alpha_0 \to \alpha_0) \to \alpha_0\}) \subseteq Cn(\{(\alpha_0 \to \alpha_0) \to \alpha_0\}).$$

As $\alpha_0 \in Cn(X_0)$ and Cn is structural, we get $\alpha_0 \in Cn(\{(\alpha_0 \to \alpha_0) \to \alpha_0\})$. But α_0 is a variable and hence $\alpha \in Cn(\{(\alpha \to \alpha) \to \alpha\})$ for every α.

Let $\alpha \in At \setminus X$ and $\alpha \in Cn_2(X) \neq S_2$. Then $(\alpha \to \alpha) \to \alpha \in Cn_2(X)$ and hence $(\alpha \to \alpha) \to \alpha \in Cn_D(X) \subseteq Cn(X)$ by $(**)$ which gives $\alpha \in Cn(X)$. $\square$

One can show that the rules r_{0i} (together with A_2) but without the rule r_1 do not axiomatize Cn_D. Namely, let $\mathfrak{M}_1$ be the matrix on $\{1,3\}$ in which 3 is designated, $\sim x = 3$ and $\triangle(x,y) = 3$ for each x,y and each $\triangle \in \{\to, +, \cdot\}$. Note that the rules r_{0i} and the axioms $Sb(A_2)$ are valid in $\mathfrak{M}_1$ whereas r_1 is not. Moreover, one can show that the consequence operation determined by these rules (and $Sb(A_2)$) coincides with $\overrightarrow{\mathfrak{M}_1} \cap \overrightarrow{\mathfrak{M}_2}$ and all its structural strengthenings are given by the following diagram:

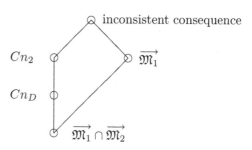

Note that $\mathfrak{M}_D$ is a submatrix of $\mathfrak{M}_1 \times \mathfrak{M}_2$ and one can also show that

$$\overrightarrow{\mathfrak{M}_D} = \overrightarrow{\mathfrak{M}_1 \times \mathfrak{M}_2}.$$

4.3 Variants

There are many variants and versions of the system (D). For instance,

Theorem 4.8. *The system* (D) *is equivalent with the following Cn-definitions:*

$$
\begin{array}{llll}
(\mathrm{D}'1) & \sim\alpha \in Cn(X) & \Leftrightarrow & Cn(X \cup \{\alpha\}) = S_2; \\
(\mathrm{D}'2) & \alpha \to \beta \in Cn(X) & \Leftrightarrow & Cn(X \cup \{\alpha, \sim\beta\}) = S_2; \\
(\mathrm{D}'3) & \alpha \cdot \beta \in Cn(X) & \Leftrightarrow & Cn(X \cup \{\alpha \to\sim \beta\}) = S_2; \\
(\mathrm{D}'4) & \alpha + \beta \in Cn(X) & \Leftrightarrow & Cn(X \cup \{\sim\alpha\cdot \sim\beta\}) = S_2.
\end{array}
$$

Proof. Note that the conditions (D′i) coincide for $i = 1, 2$ with the appropriate conditions (D). So, we assume that a consequence operation Cn fulfills (D1) and (D2) and prove that (D4) is equivalent with (D′3):

$$Cn(X \cup \{\sim\alpha\}) \cap Cn(X \cup \{\sim\beta\}) = S_2 \quad \Leftrightarrow \quad Cn(X \cup \{\alpha \to\sim \beta\}) = S_2.$$

Suppose that $Cn(X \cup \{\sim\alpha\}) = S_2$ and $Cn(X \cup \{\sim\beta\}) = S_2$. Then, by (D1), $\sim\sim\alpha \in Cn(X)$ and $\sim\sim\beta \in Cn(X)$. Since $Cn(X \cup \{\alpha \to\sim \beta, \alpha, \sim\sim\beta\}) = S_2$ by (D2) we immediately get $Cn(X \cup \{\alpha \to\sim \beta\}) = S_2$. If $Cn(X \cup \{\alpha \to\sim \beta\}) = S_2$, then $Cn(X \cup \{\sim\alpha\}) = S_2$ and $Cn(X \cup \{\sim\beta\}) = S_2$ as $\alpha \to\sim \beta \in Cn(\{\sim\alpha\}) \cap Cn(\{\sim\beta\})$.

Next, we assume that Cn fulfills (D1), (D2), (D4) and prove that the condition (D3) is equivalent with (D′4):

$$Cn(X \cup \{\sim\alpha, \sim\beta\}) = S_2 \quad \Leftrightarrow \quad Cn(X \cup \{\sim\alpha\cdot \sim\beta\}) = S_2.$$

Suppose that $Cn(X \cup \{\sim\alpha, \sim\beta\}) = S_2$. Since $\sim\alpha \in Cn(\{\sim\alpha\cdot \sim\beta\})$ and $\sim\beta \in Cn(\{\sim\alpha\cdot \sim\beta\})$ by Lemma 4.2 (ii), we also get $Cn(X \cup \{\sim\alpha\cdot \sim\beta\}) = S_2$.
Suppose that $Cn(X \cup \{\sim\alpha\cdot \sim\beta\}) = S_2$. As $Cn(\{\sim\alpha, \sim\beta, \sim\sim\alpha\}) = S_2$ and $Cn(\{\sim\alpha, \sim\beta, \sim\sim\beta\}) = S_2$, then $\sim\alpha\cdot \sim\beta \in Cn(X \cup \{\sim\alpha, \sim\beta\}) = S_2$ by (D4) and hence $Cn(X \cup \{\sim\alpha, \sim\beta\}) = S_2$. $\square$

The above theorem suggests many other sets of Cn-definitions equivalent to (D). It suffices only to take into account mutual definability of various sets of propositional connectives in classical logic. In particular, it turns out that the system (D) is equivalent with the following Cn-definitions:

$$
\begin{array}{lll}
(\text{D}''1) & \sim\alpha \in Cn(X) & \Leftrightarrow \quad Cn(X \cup \{\alpha\}) = S_2; \\
(\text{D}''2) & \alpha + \beta \in Cn(X) & \Leftrightarrow \quad Cn(X \cup \{\sim\alpha, \sim\beta\}) = S_2; \\
(\text{D}''3) & \alpha \cdot \beta \in Cn(X) & \Leftrightarrow \quad Cn(X \cup \{\sim\alpha + \sim\beta\}) = S_2; \\
(\text{D}''4) & \alpha \to \beta \in Cn(X) & \Leftrightarrow \quad Cn(X \cup \{\alpha \cdot \sim\beta\}) = S_2.
\end{array}
$$

One easily notices that the above conditions have a similar form. Namely, they all fall under the general schema:

$$
\alpha \in Cn(X) \quad \Leftrightarrow \quad Cn(X \cup \{\alpha_1, \ldots, \alpha_k\}) = S_2
$$

where $\alpha \equiv \sim (\alpha_1 \cdot \ldots \cdot \alpha_k)$ is classically valid. If α is not a variable, all conditions of the above form — if classically valid — are clearly fulfilled by the consequence operation Cn_D (see the conditions (*) and (**) in the previous section). We do not know if one can get a set of Cn-definitions of the above form which is not equivalent with the system (D). But any such set of conditions would be satisfied by the consequence operation Cn_D and hence would not be a set of Cn-definitions for the classical logic.

Thus, to get a set of Cn-definitions for logics different from Cn_D (presumably, for the classical logic), we should get rid of the above form for at least one of the conditions. Let us consider, for instance, the system (E):

$$
\begin{array}{lll}
(\text{E}1) & \sim\alpha \in Cn(X) & \Leftrightarrow \quad Cn(X \cup \{\alpha\}) = S_2; \\
(\text{E}2) & \alpha \to \beta \in Cn(X) & \Leftrightarrow \quad Cn(X \cup \{\alpha, \sim\beta\}) = S_2; \\
(\text{E}3) & \alpha + \beta \in Cn(X) & \Leftrightarrow \quad Cn(X \cup \{\sim\alpha, \sim\beta\}) = S_2; \\
(\text{E}4) & \alpha \cdot \beta \in Cn(X) & \Leftrightarrow \quad Cn(X \cup \{\alpha, \beta\}) \subseteq Cn(X).
\end{array}
$$

The above Cn-definitions are not as uniform as for the system (D) but all specified requirements are fulfilled here. Let us try to identify the logic determined by this set of Cn-definitions. First, we can show that the matrix consequence of the following 3-element matrix $\mathfrak{M}_E$, in which 3 is the only designated element, satisfies the conditions (E):

$\to$	1	2	3		$+$	1	2	3		$\cdot$	1	2	3		$\sim$	
1	3	3	3		1	1	3	3		1	1	1	1		1	3
2	1	3	3		2	3	3	3		2	1	2	2		2	1
3	1	3	3		3	3	3	3		3	1	2	3		3	1

Note that the $\{\sim, \to, +\}$ fragment of the matrix $\mathfrak{M}_E$ coincides with the same fragment of the matrix $\mathfrak{M}_D$ and the conditions (E1)–(E3) coincide with (D1)–(D3). It does not mean, however, that $\overrightarrow{\mathfrak{M}_E}$ fulfills these conditions (even if $\overrightarrow{\mathfrak{M}_D}$ does). One could show it would not be the case if the conjunction were defined in another way. It means we should prove

Lemma 4.9. *The consequence operation* $\overrightarrow{\mathfrak{M}_E}$ *fulfills the conditions* (E).

Proof. (E1): Suppose that $\sim \alpha \in \overrightarrow{\mathfrak{M}_E}(X)$. Then $h^v(X) \subseteq \{3\}$ yields $h^v(\alpha) = 1$, for every valuation v. Thus, the set $X \cup \{\alpha\}$ is not satisfiable in the matrix $\mathfrak{M}_E$ and hence $\overrightarrow{\mathfrak{M}_E}(X \cup \{\alpha\}) = S_2$.

Assume that $\overrightarrow{\mathfrak{M}_E}(X \cup \{\alpha\}) = S_2$. Then the set $X \cup \{\alpha\}$ is not satisfiable in the matrix $\mathfrak{M}_E$ and hence $h^v(X\} \subseteq \{3\}$ yields $h^v(\alpha) \in \{1,2\}$ for every valuation v. Suppose that $h^v(X) \subseteq \{3\}$ and $h^v(\alpha) \neq 3$ for some v. Let us note that the mapping $f : \{1,2,3\} \to \{1,3\}$ such that $f(1) = 1$ and $f(2) = f(3) = 3$ is a homomorphism of the algebra of the matrix $\mathfrak{M}_E$ onto the algebra of its submatrix on $\{1,3\}$. Then $X \cup \{\alpha\}$ would be satisfiable in the submatrix which would also give its satisfiability in $\mathfrak{M}_E$ — we get to a contradiction. Thus, $h^v(X) \subseteq \{3\}$ yields $h^v(\alpha) = 1$ and hence $h^v(\sim \alpha) = 3$, for every v. This means that $\sim \alpha \in \overrightarrow{\mathfrak{M}_E}(X)$.

(E2): If $\alpha \to \beta \in \overrightarrow{\mathfrak{M}_E}(X)$ and $h^v(X) \subseteq \{3\}$, then $h^v(\alpha \to \beta) = 3$ and hence $h^v(\{\alpha, \sim \beta\}) \subseteq \{3\}$ does not hold. Thus, $\overrightarrow{\mathfrak{M}_E}(X \cup \{\alpha, \sim \beta\}) = S_2$ as the set $X \cup \{\alpha, \sim \beta\}$ is not satisfiable in the matrix $\mathfrak{M}_E$.

Assume that $\overrightarrow{\mathfrak{M}_E}(X \cup \{\alpha, \sim \beta\}) = S_2$. The set $X \cup \{\alpha, \sim \beta\}$ is not satisfiable in $\mathfrak{M}_E$ and hence $h^v(X, \alpha) \subseteq \{3\}$ yields $h^v(\beta) \in \{2,3\}$, for every v. Suppose that $h^v(X) \subseteq \{3\}$ and $h^v(\alpha) = 2$ and $h^v(\beta) = 1$ for some v. Since the mapping $f: \{1,2,3\} \to \{1,3\}$ defined above is a homomorphism of the algebra of the matrix $\mathfrak{M}_E$ onto the algebra of its submatrix, the set $X \cup \{\alpha, \sim \beta\}$ would be satisfiable in the submatrix and, thus, we would get to a contradiction. This means that $(\alpha \to \beta) \in \in \overrightarrow{\mathfrak{M}_E}(X)$ as required.

(E3): If $\alpha + \beta \in \overrightarrow{\mathfrak{M}_E}(X)$ and $h^v(X) \subseteq \{3\}$, then $h^v(\alpha + \beta) = 3$ and hence $h^v(\{\sim \alpha, \sim \beta\}) \subseteq \{3\}$ does not hold. Thus, $\overrightarrow{\mathfrak{M}_E}(X, \sim \alpha, \sim \beta) = S_2$.

Assume that $\overrightarrow{\mathfrak{M}_E}(X \cup \{\sim \alpha, \sim \beta\}) = S_2$. The set $X \cup \{\sim \alpha, \sim \beta\}$ is not satisfiable in $\mathfrak{M}_E$. Suppose that $h^v(X\} \subseteq \{3\}$ and let $h^v(\alpha) \in \{2,3\}$ and $h^v(\beta) \in \{2,3\}$ for some v. Since the mapping $f : \{1,2,3\} \to \{1,3\}$ defined above is a homomorphism from the algebra of the matrix $\mathfrak{M}_E$ onto the algebra of its submatrix, the set $X \cup \{\sim \alpha, \sim \beta\}$ would be satisfiable in the submatrix and, thus, we would get to a contradiction.

(E4): Follows quite easily as the conjunction is interpreted as the minimum operator with respect to the natural ordering on the set $\{1,2,3\}$. $\qquad\square$

Using the matrix $\mathfrak{M}_E$, one easily notices that the rule r_{04} (the fragment of the modus ponens rule in which the conclusion is a conjunction) does not follow from the conditions (E). However, we can derive the following rules, instead:

$$\frac{\alpha \cdot \beta}{\alpha}, \qquad \frac{\alpha \cdot \beta}{\beta}, \qquad \frac{\alpha, \beta}{\alpha \cdot \beta}.$$

Let Cn_E be the consequence operation determined by the axiom and rules of Cn_D, except for the rule r_{04}, together with the ones above. One shows

Theorem 4.10. *The consequence operation Cn_E is the least one which fulfills the conditions* (E). *We get $Cn_E = \overrightarrow{\mathfrak{M}_E}$ and $Cn_E(\emptyset) = Z_2$. The consequence operation Cn_E is threshold, i.e.,*

$$Cn_E < Cn \in \mathrm{STRUCT} \cap \mathrm{CNS} \;\Rightarrow\; Cn = Cn_2.$$

Proof. One can show — in a standard way — that any formula is equivalent over Cn_E to a conjunction of formulas in $\{\sim, \rightarrow, +\}$. The rules and axioms of Cn_D (without r_{04}) are sufficient to show that for every α there are $\alpha_1, \ldots, \alpha_k$ which do not contain the connective $\cdot$ such that

$$\alpha \equiv \alpha_1 \cdot \ldots \cdot \alpha_k \in Cn_E(\emptyset).$$

The additional rules for $\cdot$ are essential to get

$$Cn_E(\{\alpha\}) = Cn_E(\{\alpha_1, \ldots, \alpha_k\}).$$

It means, in particular, that each structural $Cn \geq Cn_E$ is uniquely determined by its $\{\sim, \rightarrow, +\}$–fragment. Now, it suffices to make use of Theorems 4.4–4.7 (or more specifically, their versions proved for the language $\{\sim, \rightarrow, +\}$) to get similar results for the system (E). $\qquad\square$

Additional Cn-systems may be received quite easily by mixing up various Cn-definitions considered so far. It does not mean, however, that we always obtain in this way a similar body of results as above. In particular, there is no symmetry between $\cdot$ and $+$. For instance, for the system

(F1) $\sim \alpha \in Cn(X) \;\Leftrightarrow\; Cn(X \cup \{\alpha\}) = S_2;$
(F2) $\alpha \rightarrow \beta \in Cn(X) \;\Leftrightarrow\; Cn(X \cup \{\alpha, \sim \beta\}) = S_2;$
(F3) $\alpha + \beta \in Cn(X) \;\Leftrightarrow\; Cn(X \cup \{\alpha\}) \cap Cn(X \cup \{\beta\}) \subseteq Cn(X);$
(F4) $\alpha \cdot \beta \in Cn(X) \;\Leftrightarrow\; Cn(X \cup \{\alpha, \beta\}) \subseteq Cn(X),$

one can define, similarly as for the system (D), a three-element matrix $\mathfrak{M}_F$, on the set $\{1, 2, 3\}$, with 3 as the only designated element:

$\rightarrow$	1	2	3		$+$	1	2	3		$\cdot$	1	2	3		$\sim$	
1	3	3	3		1	1	2	3		1	1	1	1		1	3
2	1	3	3		2	2	2	3		2	1	2	2		2	1
3	1	3	3		3	3	3	3		3	1	2	3		3	1

It turns out that $\overrightarrow{\mathfrak{M}_F}$ fulfills the conditions (F) but this is not the least consequence operation with this property. Using the matrix one can also notice underivability of $\alpha + \sim \alpha$ in the system (F).

In none of the systems considered in this section were we able to derive the modus ponens rule (only certain of its subrules such as r_{01} etc.). However, there are systems of Cn-definitions which determine logics with the modus ponens rule. For instance, the set (H) determines the intuitionistic logic Cn_i for which r_0 is valid. Let us note, additionally, that $Cn_i(\emptyset) \neq Z_2$, nor is Cn_i a threshold logic. One can also get the modus ponens rule using the following set of Cn-definitions:

(G1) $\sim\alpha \in Cn(X) \iff Cn(X \cup \{\alpha\}) = S_2$;

(G2) $\alpha \to \beta \in Cn(X) \iff Cn(X \cup \{\sim\alpha\}) \cap Cn(X \cup \{\beta\}) = Cn(X)$;

(G3) $\alpha + \beta \in Cn(X) \iff Cn(X \cup \{\alpha\}) \cap Cn(X \cup \{\beta\}) = Cn(X)$;

(G4) $\alpha \cdot \beta \in Cn(X) \iff Cn(X \cup \{\alpha\}) \cup Cn(X \cup \{\beta\}) = Cn(X)$.

The received logic, that is the least logic fulfilling the conditions (G), is rather weak. In particular, the formula $\alpha \to \alpha$ is not valid there as shown by the following matrix, the matrix consequence of which fulfills (G):

$\to$	1	2	3		$+$	1	2	3		$\cdot$	1	2	3		$\sim$	
1	3	3	3		1	1	2	3		1	1	1	1		1	3
2	1	2	3		2	2	2	3		2	1	2	2		2	1
3	1	2	3		3	3	3	3		3	1	2	3		3	1

The Cn-definitions considered above involve set-theoretical operations. One may also say that they involve metalogical connectives (occurring in the definitions of set-theoretical operations). We conclude, therefore, that our attempts to find a set of Cn-definitions for the connectives of classical logic failed even though we accepted metalogical connectives in our conditions. In the next section we will additionally use quantification over sets of formulas to achieve our goals.

4.4   The system (I)

The following set of Cn-definitions, called the system (I), involve quantifiers ranging over subsets (or finite subsets — it does not matter) of S_2.

(I1) $\sim\alpha \in Cn(X) \iff Cn(X \cup \{\alpha\}) = S_2$;

(I2) $\alpha \to \beta \in Cn(X) \iff \forall_Y\{\forall_Z[\alpha \in Cn(X \cup Y \cup Z \cup \{\beta\})$
$\Rightarrow \alpha \in Cn(X \cup Y \cup Z)] \Rightarrow \beta \in Cn(X \cup Y)\}$;

(I3) $\alpha + \beta \in Cn(X) \iff ((\alpha \to \beta) \to \beta) \in Cn(X)$;

(I4) $\alpha \cdot \beta \in Cn(X) \iff Cn(X \cup \{\alpha,\beta\}) = Cn(X)$.

The right-hand side of the conditions (I3) should be read as

$$\forall_W\big(\forall_Y\{\forall_Z[\alpha \in Cn(X \cup W \cup Y \cup Z \cup \{\beta\}) \Rightarrow \alpha \in Cn(X \cup W \cup Y \cup Z)]$$

$$\Rightarrow \beta \in Cn(X \cup W \cup Y)\} \Rightarrow \beta \in Cn(X \cup W)\big).$$

The above is too complicated for any practical use and so we replaced it with its (less elementary but simpler) equivalent. Quantifiers are involved in our Cn-definitions and hence it is not clear if classical logic fulfills them. More specifically, it is not clear if Cn_2 fulfills (I2). In contrast to the system (D), the meet of a family of consequence operations fulfilling the conditions (I) may not fulfill these conditions. So, it is not clear whether the least consequence operation fulfilling (I) ever exists (and if it is structural and finitary).

Lemma 4.11. *The consequence operation Cn_2 fulfills the conditions* (I).

Proof. First, let us assume that the right-hand side of the condition (I2) holds for Cn_2. Let us take $Y = X \cup \{\alpha\}$ and note that $\alpha \in Cn_2(X \cup Y \cup Z)$. Hence $\beta \in Cn_2(X \cup \{\alpha\})$ which gives $(\alpha \to \beta) \in Cn_2(X)$.

Then, let us assume $(\alpha \to \beta) \in Cn_2(X)$ and suppose that for each Z:

$$\alpha \in Cn_2(X \cup Y \cup Z \cup \{\beta\}) \quad \Rightarrow \quad \alpha \in Cn_2(X \cup Y \cup Z).$$

Take $Z = \{\beta \to \alpha\}$ to get $\alpha \in Cn_2(X \cup Y \cup Z)$ and note that this suffices to get $\beta \in Cn_2(X \cup Y)$ since $(\alpha \to \beta) \in Cn_2(X)$. $\square$

Let $A^{\to +}$ denote a set of axioms for the $\{\to, +\}$ fragment of classical logic. For instance, suppose that the set consists of the axioms (1)–(6) and Peirce's law: $((\alpha \to \beta) \to \alpha) \to \alpha$ for each α, β.

Theorem 4.12. *If a consequence operation Cn fulfills the conditions* (I), *then we have $A^{\to +} \subseteq Cn(\emptyset)$ and $\beta \in Cn(\{\alpha, \alpha \to \beta\})$, for every $\alpha, \beta \in S_2$.*

Proof. Take $X = \{\alpha \to \beta\}$. Then, by (I2), one gets $\beta \in Cn(X \cup Y)$ if

$$\alpha \in Cn(X \cup Y \cup Z \cup \{\beta\}) \quad \Rightarrow \quad \alpha \in Cn(X \cup Y \cup Z)$$

for each Z. As the above holds for $Y = \{\alpha\}$, one gets $\beta \in Cn(\{\alpha, \alpha \to \beta\})$.

(1): Let us assume that

(i) $\forall_Z[\alpha \in Cn(Y \cup Z \cup \{\beta \to \alpha\}) \quad \Rightarrow \quad \alpha \in Cn(Y \cup Z)]$,

(ii) $\forall_{Z'}[\beta \in Cn(Y \cup Y' \cup Z' \cup \{\alpha\}) \quad \Rightarrow \quad \beta \in Cn(Y \cup Y' \cup Z')]$,

and show $\alpha \in Cn(Y \cup Y')$. By (i), it suffices to prove $\alpha \in Cn(Y \cup Y' \cup \{\beta \to \alpha\})$. Since we have $(\beta \to \alpha) \in Cn(Y \cup Y' \cup \{\beta \to \alpha\})$, we get by (I2),

$$\forall_{Y''}\{\forall_{Z''}[\beta \in Cn(Y \cup Y' \cup Y'' \cup Z'' \cup \{\beta \to \alpha, \alpha\})$$

$$\Rightarrow \beta \in Cn(Y, Y', Y'', Z'', \{\beta \to \alpha\})] \Rightarrow \alpha \in Cn(Y \cup Y' \cup Y'' \cup \{\beta \to \alpha\})\}.$$

Choosing suitable parameters in (ii) we get the antecedent of the above implication. Hence $\alpha \in Cn(Y \cup Y' \cup Y'' \cup \{\beta \to \alpha\})$ for each Y''.

(2): The axiom is a consequence of the remaining axioms in $A^{\to, +}$, so its derivation by use of the condition (I2) could be skipped. Let us assume that

(i) $\forall_Z[(\alpha \to (\alpha \to \beta)) \in Cn(Y \cup Z \cup \{\alpha \to \beta\}) \Rightarrow (\alpha \to (\alpha \to \beta)) \in Cn(Y \cup Z)]$,

(ii) $\forall_{Z'}[\alpha \in Cn(Y \cup Y' \cup Z \cup \beta) \Rightarrow \alpha \in Cn(Y \cup Y' \cup Z')]$, and show that $\beta \in Cn(Y \cup Y')$. We get $(\alpha \to (\alpha \to \beta)) \in Cn(\{\alpha \to \beta\})$ by (I2). Hence $(\alpha \to (\alpha \to \beta)) \in Cn(Y)$ by (i). Suppose that

$$\alpha \in Cn(Y \cup Y' \cup Z'' \cup \{\alpha \to \beta\}) \text{ for any } Z''.$$

Since $(\alpha \to \beta) \in Cn(\{\beta\})$ by (I2), we get $\alpha \in Cn(Y \cup Y' \cup Z'' \cup \{\beta\})$ and hence $\alpha \in Cn(Y \cup Y' \cup Z'')$ by (ii). Thus we have shown

$$\alpha \in Cn(Y \cup Y' \cup Z'' \cup \{\alpha \to \beta\}) \Rightarrow \alpha \in Cn(Y \cup Y' \cup Z'')$$

which gives $(\alpha \to \beta) \in Cn(Y \cup Y')$ by (I2) since $(\alpha \to (\alpha \to \beta)) \in Cn(Y)$. Using (ii), we conclude that $\beta \in Cn(Y \cup Y')$.

(3): We assume

(i) $(\forall_Z[(\alpha \to \beta) \in Cn(Y \cup Z \cup \{(\beta \to \gamma) \to (\alpha \to \gamma)\}) \Rightarrow (\alpha \to \beta) \in Cn(Y \cup Z)]$,

(ii) $\forall_{Z'}[(\beta \to \gamma) \in Cn(Y \cup Y' \cup Z' \cup \{\alpha \to \gamma\}) \Rightarrow (\beta \to \gamma) \in Cn(Y \cup Y' \cup Z')]$,

(iii) $\quad \forall_{Z''}[\alpha \in Cn(Y \cup Y' \cup Y'' \cup Z'' \cup \{\gamma\}) \Rightarrow \alpha \in Cn(Y \cup Y' \cup Y'' \cup Z'')]$,

and show $\gamma \in Cn(Y \cup Y' \cup Y'')$. Using (ii) and (I2), we easily get

$$(\alpha \to \gamma) \in Cn(Y \cup Y' \cup Y'' \cup \{(\beta \to \gamma) \to (\alpha \to \gamma)\}).$$

Hence, by (iii) and (I2), we have

(iv) $\qquad \gamma \in Cn(Y \cup Y' \cup Y'' \cup \{(\beta \to \gamma) \to (\alpha \to \gamma)\})$.

Then, by the axiom (1) and the modus ponens rule, it follows that

$$(\beta \to \gamma) \in Cn(Y \cup Y' \cup Y'' \cup \{\alpha \to \gamma\})$$

and hence, by (ii) and (I2), we obtain

$$(\beta \to \gamma) \in Cn(Y \cup Y' \cup Y'').$$

Now, according to (I2), we get $\gamma \in Cn(Y \cup Y' \cup Y'')$ as required if we show

$$\forall_W[\beta \in Cn(Y \cup Y' \cup Y'' \cup W \cup \{\gamma\}) \Rightarrow \beta \in Cn(Y \cup Y' \cup Y'' \cup W)].$$

So, let us assume that

(v) $\qquad \beta \in Cn(Y \cup Y' \cup Y'' \cup W \cup \{\gamma\})$.

Then, using the axiom (1) (together with r_0) and (iv),

$$(\alpha \to \beta) \in Cn(Y \cup Y' \cup Y'' \cup W \cup \{(\beta \to \gamma) \to (\alpha \to \gamma)\}),$$

and hence by (i) we obtain $(\alpha \to \beta) \in Cn(Y \cup Y' \cup Y'' \cup W)$. Now, by (I2), we get $\beta \in Cn(Y \cup Y \cup Y'' \cup W)$ if we show that for each W',

$$\alpha \in Cn(Y \cup Y' \cup Y'' \cup W \cup W' \cup \{\beta\}) \Rightarrow \alpha \in Cn(Y \cup Y' \cup Y'' \cup W \cup W').$$

So, suppose that $\alpha \in Cn(Y \cup Y' \cup Y'' \cup W \cup W' \cup \{\beta\})$. Then, by (v) and (iii), we get $\alpha \in Cn(Y \cup Y' \cup Y'' \cup W \cup W')$ which completes our argument.

(Peirce's law) Let us assume that

(i) $\forall_Z[((\beta \to \alpha) \to \beta) \in Cn(Y \cup Z \cup \{\beta\}) \Rightarrow ((\beta \to \alpha) \to \beta) \in Cn(Y \cup Z)]$

and show that $\beta \in Cn(Y)$. Since $((\beta \to \alpha) \to \beta) \in Cn(\{\beta\})$ by (I2), the above (i) reduces to $((\beta \to \alpha) \to \beta) \in Cn(Y)$. Thus, it suffices to show that $\beta \in Cn(\{(\beta \to \alpha) \to \beta)\})$. But $((\beta \to \alpha) \to \beta) \in Cn(\{(\beta \to \alpha) \to \beta)\})$ and hence one gets $\beta \in Cn(\{(\beta \to \alpha) \to \beta\})$ if one shows

$$\forall_{Z'}[(\beta \to \alpha) \in Cn(Z' \cup \{\beta \to \alpha) \to \beta, \beta\})$$

$$\Rightarrow (\beta \to \alpha) \in Cn(Z' \cup \{(\beta \to \alpha\ ) \to \beta\})].$$

So, let us assume $(\beta \to \alpha) \in Cn(Z' \cup \{(\beta \to \alpha) \to \beta, \beta\})$. Then by (I2) and known properties of consequence operations one gets

(ii) $(\beta \to \alpha) \in Cn(Z' \cup \{\beta\})$.

To show $(\beta \to \alpha) \in Cn(Z' \cup \{(\beta \to \alpha) \to \beta\})$ we use (I2) and hence we assume

(iii) $\forall_{Z''}[\beta \in Cn(Z' \cup Y'' \cup Z' \cup \{(\beta \to \alpha) \to \beta, \alpha\})$

$$\Rightarrow \beta \in Cn(Z' \cup Y'' \cup Z'' \cup \{(\beta \to \alpha) \to \beta\})]$$

and try to show $\alpha \in Cn(Z' \cup Y'' \cup \{(\beta \to \alpha) \to \beta\})$. Since the modus ponens rule is derivable for Cn and $(\beta \to \alpha) \in Cn(\{\alpha\})$, we get by (iii),

$$\beta \in Cn(Z' \cup Y'' \cup \{(\beta \to \alpha) \to \beta\}).$$

Then, we receive $\alpha \in Cn(Z' \cup Y'' \cup \{(\beta \to \alpha) \to \beta\})$ using (ii) and r_0.

(4)–(6): It has been shown that Cn contains the implicational fragment of classical logic. Since $\alpha + \beta$ is classically equivalent to $(\alpha \to \beta) \to \beta$ and (I3) says the same holds for Cn, we conclude that Cn contains the $\{\to, +\}$-fragment of classical logic. For instance, to show the axiom (4) we began with $(\alpha \to ((\alpha \to \beta) \to \beta)) \in Cn(\emptyset)$. By (I2), it means that

$$\forall_Y\{\forall_Z[\alpha \in Cn(Y \cup Z \cup \{(\alpha \to \beta) \to \beta\}) \Rightarrow \alpha \in Cn(Y \cup Z)]$$

$$\Rightarrow ((\alpha \to \beta) \to \beta) \in Cn(Y)\}.$$

Now, according to (I3), we can replace the formula $(\alpha \to \beta) \to \beta$ with $\alpha + \beta$ and thus we get $(\alpha \to (\alpha + \beta)) \in Cn(\emptyset)$. $\square$

Thus, each consequence operation fulfilling the conditions (I) contains the $\{\rightarrow, +\}$-fragment of classical logic together with the modus ponens. The remaining axioms are more problematic. Without difficulty one shows $(\sim \alpha \rightarrow \alpha) \rightarrow \alpha$ and $(\alpha \rightarrow \sim \alpha) \rightarrow \sim \alpha$. But $\alpha \rightarrow (\sim \alpha \rightarrow \beta)$ and $\sim\sim \alpha \rightarrow \alpha$, the axioms (10) and (12), are not derivable. More surprisingly, the axioms (7) and (8) are not derivable, either, whereas (9) as well as $\alpha \rightarrow (\beta \rightarrow \alpha \cdot \beta)$ are derivable. One could only derive the inferential versions of (7),(8) and (10):

$$\frac{\alpha, \sim \alpha}{\beta}, \qquad \frac{\alpha \cdot \beta}{\alpha}, \qquad \frac{\alpha \cdot \beta}{\beta}.$$

The above means, in particular, that the usual deduction theorem cannot be derived from (I). Instead, one can prove:

Lemma 4.13. *If a consequence operation Cn fulfills the conditions* (I), *then*

$$Cn(X \cup \{\beta\}) = Cn(X \cup \{\alpha\}) \;\Rightarrow\; (\alpha \rightarrow \beta) \in Cn(X), \quad \text{for every } X, \alpha, \beta.$$

Proof. Let us assume that $Cn(X \cup \{\beta\}) = Cn(X \cup \{\alpha\})$. We use (I2) to show that $(\alpha \rightarrow \beta) \in Cn(X)$. So, let us assume that

$$\forall_Z [\alpha \in Cn(X \cup Y \cup Z \cup \{\beta\}) \;\Rightarrow\; \alpha \in Cn(X \cup Y \cup Z)].$$

The antecedent of the implication is fulfilled as $Cn(X \cup \{\beta\}) = Cn(X \cup \{\alpha\})$. Thus, we get $\alpha \in Cn(X \cup Y \cup Z)$ for each Y, Z and hence $\alpha \in Cn(X)$. It also gives $\beta \in Cn(X)$ which was to be proved. $\qquad\square$

Corollary 4.14. *If Cn fulfills the conditions* (I), *then*

(i)	$\alpha \cdot \sim \alpha \in Cn(X)$	$\Leftrightarrow$	$\beta \cdot \sim \beta \in Cn(X)$;	
(ii)	$\sim \alpha \in Cn(X)$	$\Leftrightarrow$	$(\alpha \rightarrow \alpha \cdot \sim \alpha) \in Cn(X)$;	
(iii)	$\alpha \in Cn(X \cup \{\beta \rightarrow \alpha\})$	$\Leftrightarrow$	$((\beta \rightarrow \alpha) \rightarrow \alpha) \in Cn(X)$;	
(iv)	$\sim \alpha \in Cn(X)$	$\Rightarrow$	$(\alpha \cdot \sim \alpha \rightarrow \alpha) \in Cn(X)$;	
(v)	$(\alpha \cdot \sim \alpha \rightarrow \alpha) \in Cn(X)$	$\Rightarrow$	$(\sim\sim \alpha \rightarrow \alpha) \in Cn(X)$.	

To show that the axioms (7), (8), (10) and (12) are not derivable there one can use a 3-element matrix $\mathfrak{M}_I$. The matrix is defined, similarly as $\mathfrak{M}_D$, on $\{1, 2, 3\}$ with 3 as the only designated element and

$\rightarrow$	1	2	3
1	3	2	3
2	1	3	3
3	1	2	3

$+$	1	2	3
1	1	3	3
2	3	2	3
3	3	3	3

$\cdot$	1	2	3
1	1	1	1
2	1	2	2
3	1	2	3

$\sim$	
1	3
2	1
3	1

Theorem 4.15. *The matrix consequence operation $\overrightarrow{\mathfrak{M}_I}$ is the least consequence operation fulfilling the conditions* (I).

Proof. First, we should prove that $\overrightarrow{\mathfrak{M}_I}$ fulfills the conditions (I1)–(I4). Note that the rule r_0 is derivable for $\overrightarrow{\mathfrak{M}_I}$. It is also clear that the conditions (I3) and (I4) are fulfilled as $h^v(\alpha \cdot \beta) = 3$ iff $h^v(\alpha) = 3 = h^v(\beta)$, and $h^v(\alpha + \beta) = h^v((\alpha \to \beta) \to \beta)$ for each valuation v. Let $\mathbf{0} = p \cdot \sim p$ and $\mathbf{1} = p \to p$ for some variable p. We get $h^v(\mathbf{0}) = 1$, and $h^v(\mathbf{1}) = 3$, and $h^v(\sim \alpha) = h^v(\alpha \to \mathbf{0})$ for each valuation v. Hence $\overrightarrow{\mathfrak{M}_I}(\{\mathbf{0}\}) = S_2$ and (I1) reduces to (I2). Thus, it suffices to prove $\overrightarrow{\mathfrak{M}_I}$ fulfills (I2).

($\Rightarrow$): Let us assume that, for some X, Y, α, β and v, we have

    (i) $\alpha \to \beta \in \overrightarrow{\mathfrak{M}_I}(X)$,

    (ii) $\forall_Z [\alpha \in \overrightarrow{\mathfrak{M}_I}(X \cup Y \cup Z \cup \{\beta\}) \Rightarrow \alpha \in \overrightarrow{\mathfrak{M}_I}(X \cup Y \cup Z)]$,

    (iii) $h^v(X \cup Y) \subseteq \{3\}$.

We need to prove $h^v(\beta) = 3$. Suppose that $h^v(\beta) \in \{1, 2\}$. By (i) and (iii), we get $h^v(\alpha \to \beta) = 3$. Then $h^v(\alpha) = h^v(\beta)$. Take $Z = \{\beta \to \alpha\}$. Since we have $\alpha \in \overrightarrow{\mathfrak{M}_I}(X \cup Y \cup Z \cup \{\beta\})$ by r_0, we get $\alpha \in \overrightarrow{\mathfrak{M}_I}(X \cup Y \cup Z)$ by (ii). Hence $h^v(\alpha) = 3$, as $h^v(X \cup Y \cup Z) \subseteq \{3\}$, which contradicts our assumptions.

($\Leftarrow$): Let $h^v(X) \subseteq \{3\}$ and $h^v(\alpha \to \beta) \neq 3$ for some v. We want to show that the right-hand side of (I2) is false. We have the following possibilities:

(a) $h^v(\alpha) = 3$ and $h^v(\beta) \in \{1, 2\}$. Take $Y = \{\alpha\}$ and notice that the above (ii) is fulfilled whereas $\beta \notin \overrightarrow{\mathfrak{M}_I}(X \cup Y)$.

(b) $h^v(\alpha) = 2$ and $h^v(\beta) = 1$. This case reduces to the above one. It suffices to notice that the mapping $h \colon \{1, 2, 3\} \to \{1, 3\}$ defined by $h(1) = 1$ and $h(2) = h(3) = 3$ is an endomorphism of the algebra of the matrix. Thus we get $h(h^v(\alpha)) = 3$ and $h(h^v(\beta)) = 1$.

(c) $h^v(\alpha) = 1$ and $h^v(\beta) = 2$. Take $Y = \{\sim \alpha\}$ and note $h^v(X \cup Y) \subseteq \{3\}$ and $h^v(\beta) = 2$. Then $\beta \notin \overrightarrow{\mathfrak{M}_I}(X \cup Y)$ and hence it remains to show the condition (ii) above. Note that $\alpha \in \overrightarrow{\mathfrak{M}_I}(X \cup Y \cup Z)$ iff the set $X \cup Y \cup Z$ is not satisfiable in the matrix, which means that $h^w(X \cup Y \cup Z) \subseteq \{3\}$ holds for no w. So, to prove (ii) it suffices to show that $X \cup Y \cup Z$ is satisfiable iff $X \cup Y \cup Z \cup \{\beta\}$ is satisfiable. But if $h^w(X \cup Y \cup Z) \subseteq \{3\}$ for some w, then we also get $h(h^w(X \cup Y \cup Z) \cup \{\beta\})) \subseteq \{3\}$ for the endomorphism h defined in (b) above. Hence $X \cup Y \cup Z \cup \{\beta\}$ is satisfiable if $X \cup Y \cup Z$ is.

Suppose that Cn is a consequence operation fulfilling the conditions (I) and show that $\overrightarrow{\mathfrak{M}_I} \leqslant Cn$. So, we assume that $\alpha_0 \notin Cn(X)$ and prove that $\alpha_0 \notin \overrightarrow{\mathfrak{M}_I}(X)$. Note that we do not assume that Cn is finitary. Let X_0 be a maximal overset of X (that is $X \subseteq X_0$) with respect to the following condition:

(i) $\alpha_0 \notin Cn(Z)$ for each finite $Z \subseteq X_0$.

It follows from the definition of X_0 and Corollary 4.14 that

(ii) for each $\beta \notin X_0$ there is a finite set $Z \subseteq X_0$ such that $\alpha_0 \in Cn(Z \cup \{\beta\})$,

(iii) either $(\beta \rightarrow \alpha_0) \in X_0$, or $((\beta \rightarrow \alpha_0) \rightarrow \alpha_0) \in X_0$, for every β.

Let us prove

(iv) $(\beta \rightarrow \alpha_0) \in X_0 \Rightarrow (\alpha_0 \rightarrow \beta) \in X_0$, for every β.

Let $(\beta \rightarrow \alpha_0) \in X_0$ and $(\alpha_0 \rightarrow \beta) \notin X_0$. Then $\alpha_0 \in Cn(Z_0 \cup \{\alpha_0 \rightarrow \beta\})$ for some finite set $Z_0 \subseteq X_0$, by (2). We can assume that $(\beta \rightarrow \alpha_0) \in Z_0$. Let us show, using (I2), that $((\alpha_0 \rightarrow \beta) \rightarrow \alpha_0) \in Cn(Z_0)$. So, suppose that

$$\forall_Z[(\alpha_0 \rightarrow \beta) \in Cn(Z_0 \cup Y \cup Z \cup \{\alpha_0\}) \Rightarrow (\alpha_0 \rightarrow \beta) \in Cn(Z_0 \cup Y \cup Z)]$$

and show that $\alpha_0 \in Cn(Z_0 \cup Y)$. Since we have $\alpha_0 \in Cn(Z_0 \cup \{\alpha_0 \rightarrow \beta\})$ and $(\alpha_0 \rightarrow \beta) \in Cn(\{\beta\})$, the above implication gives us

$$\forall_Z[\beta \in Cn(Z_0 \cup Y \cup Z \cup \{\alpha_0\}) \Rightarrow \beta \in Cn(Z_0 \cup Y \cup Z)].$$

Using (I2) and $(\beta \rightarrow \alpha_0) \in Z_0 \subseteq Cn(Z_0)$, we get $\alpha_0 \in Cn(Z_0 \cup Y)$ as required. Then, by Peirce's law, $\alpha_0 \in Cn(Z_0)$ which contradicts (i).

Suppose that $(\mathbf{0} \rightarrow \alpha_0) \in X_0$. Then $(\alpha_0 \rightarrow \mathbf{0}) \in X_0$ by (iv). Thus, for each $\beta \notin X_0$, we get $Cn(X_0 \cup \{\beta\}) = Cn(X_0 \cup \{\mathbf{0}\})$ and hence $\beta \equiv \mathbf{0} \in X_0$, by Lemma 4.13. It means that the set of all formulas is divided into two disjoint classes: formulas equivalent to $\mathbf{1}$, and those equivalent to $\mathbf{0}$ (over X_0). The division defines in a natural way a valuation in the two-element matrix $\mathfrak{M}_2$. Thus, we get $\alpha_0 \notin \overrightarrow{\mathfrak{M}_2}(X_0)$ and hence $\alpha_0 \notin \overrightarrow{\mathfrak{M}_I}(X)$ as $\overrightarrow{\mathfrak{M}_I}(X) \subseteq \overrightarrow{\mathfrak{M}_2}(X_0)$.

Suppose that $(\mathbf{0} \rightarrow \alpha_0) \notin X_0$. Then, by (iii), $((\mathbf{0} \rightarrow \alpha_0) \rightarrow \alpha_0) \in X_0$ and hence, as Cn fulfills the conditions (I) — see Theorem 4.12, we get

(v) $((\alpha_0 \rightarrow \mathbf{0}) \rightarrow \mathbf{0})) \in X_0$.

Let X_1 be a maximal overset of X_0 which does not contain the formula α_0 and which is closed under the following rules:

$$\frac{\phi, \phi \rightarrow \psi}{\psi}, \qquad \frac{\phi \cdot \sim \phi}{\psi}.$$

(vi) $\beta \notin X_1 \quad \Leftrightarrow \quad (\beta \rightarrow \alpha_0) \in X_1$ or $(\beta \rightarrow \mathbf{0}) \in X_1$.

Indeed, if $\beta \notin X_1$, one can derive α_0 using the above rules from $X_1 \cup \{\beta\}$. If only the modus ponens occurs in the proof we get $(\beta \rightarrow \alpha_0) \in X_1$ by the classical deduction theorem. If the second rule occurs, we get an r_0-proof of $\mathbf{0}$ and hence $(\beta \rightarrow \mathbf{0}) \in X_1$ by the deduction theorem.

Next, let us show

(vii) $(\beta \rightarrow \mathbf{0}) \in X_1 \quad \Rightarrow \quad (\mathbf{0} \rightarrow \beta) \in X_1$.

Let $(\beta \rightarrow \mathbf{0}) \in X_1$ and $(\mathbf{0} \rightarrow \beta) \notin X_1$. We have $((\mathbf{0} \rightarrow \beta) \rightarrow \mathbf{0}) \notin X_1$ (as otherwise $\mathbf{0} \in X_1$ by Peirce's law) and hence $((\mathbf{0} \rightarrow \beta) \rightarrow \alpha_0) \in X_1$ by (vi). Thus, by (iii), we get $((\mathbf{0} \rightarrow \beta) \rightarrow \alpha_0) \in X_0$ and hence $(\beta \rightarrow \alpha_0) \in X_0$. Then

$(\alpha_0 \to \beta) \in X_0$ by (vi). Since $X_0 \subseteq X_1$ and $(\beta \to \mathbf{0}) \in X_1$, we get $(\alpha_0 \to \mathbf{0}) \in X_1$ which contradicts (v).

If $(\beta \to \alpha_0) \in X_1$, then $(\beta \to \alpha_0) \in X_0$ by (iii) and hence $(\beta \equiv \alpha_0) \in X_1$ by (iv). If $(\beta \to \mathbf{0}) \in X_1$, then $(\beta \equiv \mathbf{0}) \in X_1$ by (vii). Thus, by (vi), the set of formulas can be divided into three disjoint classes: formulas equivalent over X_1 to $\mathbf{1}$, $\mathbf{0}$ or α_0, respectively. Moreover, the following formulas belong to X_1:

$\mathbf{0} \to \mathbf{0}$	$\equiv$	$\mathbf{1}$	$\mathbf{0} \to \alpha_0$	$\equiv$	α_0	$\mathbf{0} \to \mathbf{1}$	$\equiv$	$\mathbf{1}$
$\alpha_0 \to \mathbf{0}$	$\equiv$	$\mathbf{0}$	$\alpha_0 \to \alpha_0$	$\equiv$	$\mathbf{1}$	$\alpha_0 \to \mathbf{1}$	$\equiv$	$\mathbf{1}$
$\mathbf{1} \to \mathbf{0}$	$\equiv$	$\mathbf{0}$	$\mathbf{1} \to \alpha_0$	$\equiv$	α_0	$\mathbf{1} \to \mathbf{1}$	$\equiv$	$\mathbf{1}$

$\mathbf{0} \cdot \mathbf{0}$	$\equiv$	$\mathbf{0}$	$\mathbf{0} \cdot \alpha_0$	$\equiv$	$\mathbf{0}$	$\mathbf{0} \cdot \mathbf{1}$	$\equiv$	$\mathbf{0}$
$\alpha_0 \cdot \mathbf{0}$	$\equiv$	$\mathbf{0}$	$\alpha_0 \cdot \alpha_0$	$\equiv$	α_0	$\alpha_0 \cdot \mathbf{1}$	$\equiv$	α_0
$\mathbf{1} \cdot \mathbf{0}$	$\equiv$	$\mathbf{0}$	$\mathbf{1} \cdot \alpha_0$	$\equiv$	α_0	$\mathbf{1} \cdot \mathbf{1}$	$\equiv$	$\mathbf{1}$

Similar equivalences can be shown for the operators $+$ and $\sim$, as using the condition (I) one easily derives $\alpha + \beta \equiv (\alpha \to \beta) \to \beta$ and $\sim \alpha \equiv (\alpha \to \mathbf{0})$. It means that the quotient algebra is isomorphic to the matrix $\mathfrak{M}_I$ (where $\mathbf{0}$ corresponds to 1, and $\mathbf{1}$ corresponds to 3). One also gets a valuation $v \colon At \to \{1, 2, 3\}$ such that $h^v(X_1) \subseteq \{3\}$ and $h^v(\alpha_0) = 2$. So $\alpha_0 \notin \overrightarrow{\mathfrak{M}_I}(X_1)$.   □

Notice that the classical two-element matrix $\mathfrak{M}_2$ is (isomorphic to) a submatrix of $\mathfrak{M}_I$. Hence, $\overrightarrow{\mathfrak{M}_I} \leqslant Cn_2$. As $\overrightarrow{\mathfrak{M}_I}(\emptyset)$ is a proper subset of Z_2, we get $\overrightarrow{\mathfrak{M}_I} < Cn_2$. Thus, the conditions (I1)–(I4) do not determine the classical logic Cn_2, only a weaker system. Nevertheless, the two logics are very close as we can prove that $\overrightarrow{\mathfrak{M}_I}$ is a threshold logic:

Theorem 4.16. *If Cn is a structural and consistent consequence operation and $Cn > \overrightarrow{\mathfrak{M}_I}$, then $Cn = Cn_2$.*

Proof. Suppose that $Cn \in \mathrm{STRUCT} \cap \mathrm{CNS}$ and $Cn \geq \overrightarrow{\mathfrak{M}_I}$. Let q be a fixed variable and denote $q \cdot \sim q$ by $\mathbf{0}$, and $q \to q$ by $\mathbf{1}$. If $\mathbf{0} \to p \in Cn(\emptyset)$, we get $Sb(A_2) \subseteq Cn(\emptyset)$ by Theorems 4.12–4.14 and hence $Cn = Cn_2$. So, let us assume that $(\mathbf{0} \to p) \notin Cn(\emptyset)$.

We consider the 3-element subalgebra of the Lindenbaum–Tarski algebra for Cn determined by: $\mathbf{0}$, $\mathbf{0} \to p$, $\mathbf{1}$. Using (I1)–(I4) one easily shows:

$\mathbf{0} \to \mathbf{0}$	$\equiv$	$\mathbf{1}$	$\mathbf{0} \to (\mathbf{0} \to p)$	$\equiv$	$\mathbf{0} \to p$	$\mathbf{0} \to \mathbf{1}$	$\equiv$	$\mathbf{1}$
$(\mathbf{0} \to p) \to \mathbf{0}$	$\equiv$	$\mathbf{0}$	$(\mathbf{0} \to p) \to (\mathbf{0} \to p)$	$\equiv$	$\mathbf{1}$	$(\mathbf{0} \to p) \to \mathbf{1}$	$\equiv$	$\mathbf{1}$
$\mathbf{1} \to \mathbf{0}$	$\equiv$	$\mathbf{0}$	$\mathbf{1} \to (\mathbf{0} \to p)$	$\equiv$	$\mathbf{0} \to p$	$\mathbf{1} \to \mathbf{1}$	$\equiv$	$\mathbf{1}$

$\mathbf{0} \cdot \mathbf{0}$	$\equiv$	$\mathbf{0}$	$\mathbf{0} \cdot (\mathbf{0} \to p)$	$\equiv$	$\mathbf{0}$	$\mathbf{0} \cdot \mathbf{1}$	$\equiv$	$\mathbf{0}$
$(\mathbf{0} \to p) \cdot \mathbf{0}$	$\equiv$	$\mathbf{0}$	$(\mathbf{0} \to p) \cdot (\mathbf{0} \to p)$	$\equiv$	$\mathbf{0} \to p$	$(\mathbf{0} \to p) \cdot \mathbf{1}$	$\equiv$	$\mathbf{0} \to p$
$\mathbf{1} \cdot \mathbf{0}$	$\equiv$	$\mathbf{0}$	$\mathbf{1} \cdot (\mathbf{0} \to p)$	$\equiv$	$\mathbf{0} \to p$	$\mathbf{1} \cdot \mathbf{1}$	$\equiv$	$\mathbf{1}$

Similar tables can be given for $+$ and $\sim$. We conclude that a subalgebra of the Lindenbaum–Tarski algebra is isomorphic to $\mathfrak{M}_I$ and hence $Cn \leqslant \overrightarrow{\mathfrak{M}_I}$.   □

One can axiomatize the consequence operation $\overrightarrow{\mathfrak{M}_I}$ using the $\overrightarrow{\mathfrak{M}_I}$ derivable formulas from A_2, Peirce's law, the inferential versions of the axioms (7), (8) and (10) (mentioned below Theorem 4.12), and one of the rules

$$\frac{\sim (\alpha \to \beta)}{\alpha + \beta}, \qquad \frac{\sim \alpha}{\alpha \cdot \sim \alpha \to \alpha}, \qquad \frac{\sim \beta, \sim\sim \alpha}{\alpha + \beta}.$$

Let us observe that the axiom (12) is not derivable in the inferential form, either. The above rules make our axiomatization of $\overrightarrow{\mathfrak{M}_I}$ useless for any practical purpose. One can show that they cannot be omitted using the following 4-element matrix on $\{0, 1, 2, 3\}$, with 3 as the only designated element, in which implication and disjunction are defined as in the 4-element Boolean algebra

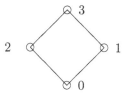

and negation is defined by $x \to 1$ (not as $x \to 0$ which is usual) and conjunction is determined by the usual linear ordering on $\{0, 1, 2, 3\}$ (not the Boolean order). Note that $\overrightarrow{\mathfrak{M}_I}$ is a submatrix of the above 4-element matrix. All axioms (1)–(11) (or corresponding rules) are valid in the matrix, but the additional rules are not.

4.5 Classical logic

Our attempts to discover a set of Cn-definitions for the classical logic failed. We conjecture that such a set does not exist. We would be successful if the idea of Cn-definability were violated somehow. For instance, the conditions (D)+(H2) characterize the classical logic (as the least consequence fulfilling them) but we get two Cn-definitions for implication there. Below we make several similar approaches in each of which the basic idea of Cn-definability is modified or Cn-definitions are extended with additional conditions.

First, we can claim that the conditions (D) concern (complex) formulas rather than propositional operators. Hence, on the basis of Corollary 4.5, there remains to extend these conditions with one which characterizes derivability of variables:

(D0) $\qquad p \in Cn(X) \Leftrightarrow Cn(X \cup \{\sim p\}) = S_2$ for every $p \in At, X \subseteq S_2$.

Corollary 4.17. Cn_2 *is the least consequence operation fulfilling the conditions* (D0)–(D4). Cn_2 *is also the only structural and consistent consequence operation fulfilling these conditions.*

Another possibility is to introduce an additional connective, assertion, denoted by $\iota\alpha$, and add to (D1)–(D4) the condition

(D5) $\qquad \iota\alpha \in Cn(X) \Leftrightarrow Cn(X \cup \{\sim \alpha\}) = S_2$.

Since $(\alpha \equiv \iota\alpha) \in Cn(\emptyset)$, we can identify the operator ι with the usual assertion. Then there remains to identify formulas with their assertions. In fact, it suffices only to identify p with ιp for each variable p to get the classical logic. We omit details as both approaches are quite similar.

Now, we consider certain contextual Cn-definitions of connectives. It turns out they are stronger than explicit Cn-definitions. Let us prove

Theorem 4.18. *Cn_2 is the least consequence operation fulfilling the conditions* (K1)–(K4) *below. Cn_2 is also the only structural and consistent consequence operation which fulfills these conditions.*

$$
\begin{aligned}
&(K1) &&Cn(X \cup \{\sim\alpha\}) = S_2 &&\Leftrightarrow &&Cn(X \cup \{\alpha\}) \subseteq Cn(X);\\
&(K2) &&Cn(X \cup \{\alpha \to \beta\}) = S_2 &&\Leftrightarrow &&Cn(X \cup \{\alpha, \sim\beta\}) \subseteq Cn(X);\\
&(K3) &&Cn(X \cup \{\alpha + \beta\}) = S_2 &&\Leftrightarrow &&Cn(X \cup \{\sim\alpha, \sim\beta\}) \subseteq Cn(X);\\
&(K4) &&Cn(X \cup \{\alpha \cdot \beta\}) = S_2 &&\Leftrightarrow &&Cn(X \cup \{\sim\alpha + \sim\beta\}) \subseteq Cn(X).
\end{aligned}
$$

Proof. Suppose that Cn is a consequence operation fulfilling (K1)–(K4). We need to show — see Corollary 4.5 — that Cn fulfills (D1)–(D4) and the modus ponens rule r_0 is a derivable rule of Cn.

(D1): Since $Cn(\{\alpha, \alpha\}) \subseteq Cn(\{\alpha\})$, we get $Cn(\{\alpha, \sim\alpha\}) = S_2$ by (K1). So, if $\sim\alpha \in Cn(X)$, then $Cn(X \cup \{\alpha\}) = S_2$. Suppose on the other hand that $Cn(X \cup \{\alpha\}) = S_2$. Since $Cn(\{\sim\sim\alpha, \sim\alpha\}) = S_2$, we get $\alpha \in Cn(\{\sim\sim\alpha\})$ by (K1) and hence $Cn(X \cup \{\sim\sim\alpha\}) = S_2$. This means that $\sim\alpha \in Cn(X)$.

(D2): We get $Cn(\{\alpha \to \beta, \alpha, \sim\beta\}) = S_2$ by (K2). So, if $\alpha \to \beta \in Cn(X)$, then $Cn(X \cup \{\alpha, \sim\beta\}) = S_2$. Suppose that $Cn(X \cup \{\alpha, \sim\beta\}) = S_2$. Since we have $Cn(\{\alpha \to \beta, \sim(\alpha \to \beta)\}) = S_2$, then $\alpha, \sim\beta \in Cn(\{\sim(\alpha \to \beta)\})$ by (K2) and hence $Cn(X \cup \{\sim(\alpha \to \beta)\}) = S_2$ which gives $\alpha \to \beta \in Cn(X)$ by (K1).

(D3): We get $Cn(\alpha + \beta, \sim\alpha, \sim\beta) = S_2$ by (K3). So, if $\alpha + \beta \in Cn(X)$, then $Cn(X \cup \{\sim\alpha, \sim\beta\}) = S_2$. Suppose that $Cn(X \cup \{\sim\alpha, \sim\beta\}) = S_2$. Since we have $Cn(\{\alpha + \beta, \sim(\alpha + \beta)\}) = S_2$, then $\sim\alpha, \sim\beta \in Cn(\{\sim(\alpha \to \beta)\})$ by (K3) and hence $Cn(X \cup \{\sim(\alpha + \beta)\}) = S_2$ which gives $\alpha + \beta \in Cn(X)$ by (K1).

(D4): We get $Cn(\{\alpha \cdot \beta, \sim\alpha + \sim\beta\}) = S_2$ by (K4). So, if $\alpha \cdot \beta \in Cn(X)$, then $Cn(X \cup \{\sim\alpha + \sim\beta\}) = S_2$ and hence $Cn(X \cup \{\sim\alpha\}) \cap Cn(X \cup \{\sim\beta\}) = S_2$ by (K3). Suppose that $Cn(X \cup \{\sim\alpha\}) \cap Cn(X \cup \{\sim\beta\}) = S_2$. Then $\alpha, \beta \in Cn(X)$ by (K1). We get $Cn(\{\alpha, \beta, \sim\alpha + \sim\beta\}) = S_2$ by (K3) and (D1). Since we get $\sim\alpha + \sim\beta \in Cn(\{\sim(\alpha \cdot \beta)\})$ by (K4), we have $Cn(\{\alpha, \beta, \sim(\alpha \cdot \beta)\}) = S_2$ which gives $\alpha \cdot \beta \in Cn(X)$ by (K1).

(r_0): We have $Cn(\{\alpha \to \beta, \alpha, \sim\beta\}) = S_2$ on the basis of (K2) and hence we get $\beta \in Cn(\{\alpha \to \beta, \alpha\})$ by (K1). □

The above contextual Cn-definitions can also be expressed in terms of sequential consequence relations. So, one can say that the connectives of classical logic can be characterized by use of certain Cn-conditions if one extends the concept of a consequence relation to a multiple consequence relation.

The above contextual Cn-definitions are close to the usual semantical characterization of the classical logic. Every valuation $v: At \to \{0, 1\}$ corresponds to a consistent consequence operation fulfilling the conditions (N1)–(N4) below. The consequence Cn_2 does not fulfill these conditions but it is the intersection of all (N)-systems:

$$
\begin{aligned}
&\text{(N1)} & \sim\alpha \in Cn(X) &\Leftrightarrow& \alpha \notin Cn(X); \\
&\text{(N2)} & \alpha \to \beta \in Cn(X) &\Leftrightarrow& \text{if } \alpha \in Cn(X) \text{ then } \beta \in Cn(X); \\
&\text{(N3)} & \alpha + \beta \in Cn(X) &\Leftrightarrow& \alpha \in Cn(X) \text{ or } \beta \in Cn(X); \\
&\text{(N4)} & \alpha \cdot \beta \in Cn(X) &\Leftrightarrow& \alpha \in Cn(X) \text{ and } \beta \in Cn(X).
\end{aligned}
$$

Let us also recall Theorems 4.7, 4.10 and 4.16. According to them any additional requirement (in addition to (D), (E) or (I)) together with the structurality of Cn gives us the classical system. In particular,

Corollary 4.19. *If Cn is a structural, consistent and structurally complete consequence operation fulfilling the conditions* (D)*, then $Cn = Cn_2$.*

One should keep in mind, however, that $\overrightarrow{\mathfrak{M}_I}$ is structurally complete — in contrast to $\overrightarrow{\mathfrak{M}_D}$ or $\overrightarrow{\mathfrak{M}_E}$ — and hence the counterpart of Corollary 4.19 does hold, as well, for the system (E) but it does not for the system (I).

The technical reason for which our attempts fail is, perhaps, the fact that the negation $\sim$ in Cn-systems is too weak. Note that the conditions (D1), and (E1), and (I1) coincide and it is difficult to imagine another Cn-definition for negation. We would succeed if we replaced, for instance, the condition (D1) in the system (D) with the following one (which is not a Cn-definition):

$$
\alpha \in Cn(X) \quad \Leftrightarrow \quad Cn(X \cup \{\sim\alpha\}) = S_2.
$$

Using (D1) one can reduce the above condition to a connective-free form. Thus, if we extend the general concept of consequence operation (defined originally by the condition (T0)) with the condition (M1) given below, we obtain a set of Cn-definitions for classical logic:

Theorem 4.20. *Cn_2 is the least operation fulfilling the following conditions:*

$$
\begin{aligned}
&\text{(M0)} & &\forall_Y[X \subseteq Cn(X) \Rightarrow Y \cup Cn(X) \subseteq Cn(Y)]; \\
&\text{(M1)} & &\forall_Y[Cn(X \cup Y \cup \{\alpha\}) = S_2 \Rightarrow Cn(X \cup Y) = S_2] \Rightarrow \alpha \in Cn(X); \\
&\text{(M2)} & &\sim\alpha \in Cn(X) \Leftrightarrow \forall_Y[\alpha \in Cn(X \cup Y) \Rightarrow Cn(X \cup Y) = S_2]; \\
&\text{(M3)} & &\alpha \to \beta \in Cn(X) \Leftrightarrow \forall_Y[Cn(X \cup Y \cup \{\beta\}) = S_2 \Rightarrow Cn(X \cup Y \cup \{\alpha\}) = S_2]; \\
&\text{(M4)} & &\alpha \cdot \beta \in Cn(X) \Leftrightarrow \forall_Y[Cn(X \cup Y \cup \{\alpha, \beta\}) = S_2 \Rightarrow Cn(X \cup Y) = S_2]; \\
&\text{(M5)} & &\alpha + \beta \in Cn(X) \Leftrightarrow \forall_Y[Cn(X \cup Y \cup \{\alpha\}) \cap Cn(X \cup Y \cup \{\beta\}) = S_2 \\
& & &\qquad\qquad \Rightarrow Cn(X \cup Y) = S_2].
\end{aligned}
$$

We denote the above set of conditions by (M). Note that (M0) coincides with (T0) and (M2)–(M5) are equivalent over (T0) to (D1)–(D4). The additional (with respect to the system (D)) condition (M1) is connective-free. We may claim that (T0)+(M1) narrows the general concept of a consequence operation. Thus, Theorem 4.20 says that Cn-definitions of the classical propositional connectives are possible if one changes the general concept of a consequence operation. Let us add that in place of (M1) one can also use other general properties of consequence operations, and instead of (M2)–(M5) one could take the conditions (I), or any other set of Cn-definitions considered in our paper.

Cn-definability seems to be well-suited for definability of propositional connectives. It works, at least, for intuitionistic logic. However, it turns out that we can get in this way threshold logics (the systems (D), (E) and (I)) but not the classical one. The classical logic can be reached if we modify Cn-definability (the system (K)) or change the general concept of a consequence operation (the system (M)). The problem arises how to prove formally that there is no set of Cn-definitions for the classical connectives without extending the concept of a consequence operation. We did not manage to get such a result. We were only able to prove, see section 2, that the connectives of classical logic do not have Cn-definitions of the form:

$$\alpha \in Cn(X) \quad \Leftrightarrow \quad Cn(X \cup \{\alpha_1, \dots, \alpha_k\}) = S_2.$$

Any formal proof of the negative result mentioned above would require a precise definition of Cn-definability. We have not specified a precise variant of the concept (it means we have not specified metalogical means allowed in Cn-definitions) but it seems to us that a language with metaconnectives and quantification over sets (or finite sets) of formulas is the proper one for this purpose. This variant is sufficiently wide to cover all Cn-systems considered in our book. Note that Cn-definitions usually appear in a certain order: negation is defined first, then one defines implication (or disjunction) and the conjunction at the end. These dependencies disappear, and each connective is defined independently, if one introduces quantifiers over sets of formulas.

Appendix A
The fundamental metatheorem for the classical propositional logic

The main result of this section, see Metatheorem A.4, describes connections between the theory of Boolean algebras and classical propositional logic.

We assume now that At is an infinite set of propositional variables which means that At need not be countable. Recall that $\alpha \in FC_2(X)$, for $X \subseteq S_2$ and $\alpha \in S_2$ iff in each Boolean algebra $\mathscr{B}$ we have $h^v(\alpha) \in F\big(h^v(X)\big)$ for all $v \colon At \to B$ where $F\big(h^v(X)\big)$ is the filter generated by the set $h^v(X)$. The operation FC_2 coincides, see Theorem 2.90, with the consequence operation determined by the classical logic:

$$FC_2(X) = Cn\big(R_0, Sb(A_2) \cup X\big), \qquad \text{for each } X \subseteq S_2.$$

It is also worth noticing that the above equality can be proved effectively.

The proof of the announced metatheorem will be technically facilitated by some facts concerning Lindenbaum–Tarski algebras. The relation $\sim_X$ (for each $X \subseteq S_2$) is defined in the following way: $\alpha \sim_X \beta$ iff $(\alpha \equiv \beta) \in FC_2(X)$. This relation is a congruence relation on the algebra of the language $\mathscr{S}_2$, and the quotient algebra

$$\mathscr{S}_2/\sim_X = \langle S_2/\sim_X, \dot{\to}, \cup, \cap, - \rangle,$$

with the relation $\leqslant_X$ defined by $[\alpha] \leqslant_X [\beta]$ iff $\alpha \to \beta \in FC_2(X)$, is a Boolean algebra in which the following holds:

$$[\alpha + \beta] = [\alpha] \cup [\beta]$$
$$[\alpha \cdot \beta] = [\alpha] \cap [\beta]$$
$$[\alpha \to \beta] = [\alpha] \dot{\to} [\beta]$$
$$[\sim \alpha] = -[\alpha]$$
$$[\alpha] = 1_X \Leftrightarrow \alpha \in FC_2(X)$$
$$[\alpha] = 0_X \Leftrightarrow \sim \alpha \in FC_2(X).$$

This is the so-called Lindenbaum–Tarski algebra determined by the set X (or by $FC_2(X)$) and is symbolized by $\mathscr{L}_2(X)$.

Lemma A.1. *Let H be a proper filter in a Boolean algebra $\mathscr{B}$ and let $h\colon \mathscr{S}_2 \to \mathscr{B}$ be a homomorphism. Then $h^{-1}(H)$ is an FC_2-consistent and FC_2-closed set, i.e.,*

$$h^{-1}(H) = FC_2\big(h^{-1}(H)\big) \neq S_2.$$

Proof. Assume $\alpha \in FC_2\big(h^{-1}(H)\big)$. Hence $h(\alpha) \in F\big(hh^{-1}(H)\big) \subseteq F(H) = H$ which gives $\alpha \in h^{-1}(H)$. Since H is a proper filter, we have $h^{-1}(H) \neq S_2$. $\square$

The next theorem is due to Łoś ([62], 1954).

Lemma A.2. *Each Boolean algebra (of the power $\leqslant \overline{\overline{At}}$) is isomorphic with some Lindenbaum–Tarski algebra $\mathscr{L}_2(X)$.*

Proof. Let $\mathscr{B}$ be a Boolean algebra such that $\overline{\overline{B}} \leqslant \overline{\overline{At}}$ and let $v\colon At \to B$ be a valuation from At onto the set B. Then the homomorphism h^v maps $\mathscr{S}_2$ onto the algebra $\mathscr{B}$. Therefore, by Lemma 1.5, $\mathscr{B}$ will be isomorphic with the quotient algebra $\mathscr{S}_2 / \approx$, where the congruence-relation is defined in the following way:

$$\alpha \approx \beta \qquad \Leftrightarrow \qquad h^v(\alpha) = h^v(\beta) \qquad \text{for } \alpha, \beta \in S_2.$$

In order to complete the proof it suffices to show that $\approx$ is a congruence relation determined by the set $X = \{\alpha : h^v(\alpha) = 1_{\mathscr{B}}\}$, i.e., $\alpha \approx \beta$ iff $\alpha \equiv \beta \in FC_2(X)$. If $\alpha \approx \beta$, that is if $h^v(\alpha) = h^v(\beta)$, then $h^v(\alpha \equiv \beta) = 1_{\mathscr{B}}$, and hence $\alpha \equiv \beta \in X$. If, on the other hand, $\alpha \equiv \beta \in FC_2(X)$, then $\alpha \equiv \beta \in X$ (by Lemma A.1, the set X is FC_2-closed) and consequently $h^v(\alpha) = h^v(\beta)$ which means that $\alpha \approx \beta$. Thus $\mathscr{S}_2 / \approx$ is a Lindenbaum–Tarski algebra isomorphic with $\mathscr{B}$. $\square$

Lemma A.3. *For each $X \subseteq S_2$ and each $\alpha \in S_2$,*

$$FC_2(X \cup \{\sim \alpha\}) \neq S_2 \qquad \Leftrightarrow \qquad \alpha \notin FC_2(X).$$

This is the well-known property of the classical propositional logic, see Corollary 1.67. Let us recall that $\mathscr{B}_2$ is the two-element Boolean algebra defined on the set $\{0, 1\}$. We also recall that ZF is the standard system of set theory *(Zermelo–Fraenkel set theory)*. Derivability in ZF, sometimes called *effective*, means derivability without using the Axiom of Choice.

Metatheorem A.4. *The following statements are equivalent in ZF (the validity of (ii)–(iv) is understood as validity for every language, i.e., for language of any infinite cardinality):*

(i) *Each Boolean algebra is isomorphic with some field of sets;*

(ii) $FC_2(X) = \overrightarrow{\mathscr{B}_2}(X)$ *for each $X \subseteq S_2$.*

(iii) *For each $X \subseteq S_2$: if $FC_2(X) \neq S_2$, then there is a valuation $v\colon At \to \{0, 1\}$ such that $h^v(X) \subseteq \{1\}$.*

(iv) *For each $X \subseteq S_2$ and $\alpha \in S_2$: if $\left(h^e(X) \subseteq FC_2(\emptyset) \Rightarrow h^e(\alpha) \in FC_2(\emptyset)\right)$ for all $e\colon At \to S_2$, then $\alpha \in FC_2(X)$.*

(v) *For each $X \subseteq S_2$: if $FC_2(X) \neq S_2$, then there is $Y \subseteq S_2$ such that $X \subseteq Y = FC_2(Y) \neq S_2$ and $\alpha \in Y$ or $\sim \alpha \in Y$ for all $\alpha \in S_2$.*

(vi) *Each Lindenbaum–Tarski algebra $\mathscr{L}_2(X)$ is isomorphic with some field of sets.*

Proof. (i)$\Rightarrow$(ii): Assume that $\alpha \notin FC_2(X)$ and let $\mathscr{L}_2(X)$ be the Lindenbaum–Tarski algebra determined by the set X. Then $\mathscr{L}_2(X)$ is a non-degenerate Boolean algebra and therefore, according to (i), $\mathscr{L}_2(X)$ is isomorphic with some field of sets $\mathscr{B}$. The unit element $1_{\mathscr{B}}$ is a non-empty set since the algebra $\mathscr{L}_2(X)$ contains at least two different elements. $\mathscr{L}_2(X)$ and $\mathscr{B}$ are isomorphic, then there exists a homomorphism (natural) $h\colon \mathscr{S}_2 \to \mathscr{B}$ such that $h(X) \subseteq \{1_{\mathscr{B}}\}$ and $h(\alpha) \neq 1_{\mathscr{B}}$. Let then $a \in 1_{\mathscr{B}}$ and $a \notin h(\alpha)$. Define the valuation $v\colon At \to \{0,1\}$ in the following way: $v(\gamma) = 1 \Leftrightarrow a \in h(\gamma)$ for all $\gamma \in At$. It can be easily seen that the mapping v extends to a homomorphism $h^v\colon \mathscr{S}_2 \to \mathscr{B}_2$ such that

$$h^v(\phi) = 1 \quad \Leftrightarrow \quad a \in h(\phi), \qquad \text{for all } \phi \in S_2.$$

The condition $h^v(X) \subseteq \{1\}$ is true because, for each $\beta \in X$, we have $a \in 1_{\mathscr{B}} = h(\beta)$ and hence $h^v(\beta) = 1$. What is more, $h^v(\alpha) \neq 1$. So it has been shown that $\alpha \notin \overrightarrow{\mathscr{B}}_2(X)$. Consequently the inclusion $\overrightarrow{\mathscr{B}}_2(X) \subseteq FC_2(X)$ holds. The reverse inclusion is obvious.

(ii)$\Rightarrow$(iii): Immediately.

(iii)$\Rightarrow$(iv): Assume that

$$(\star) \qquad h^e(X) \subseteq FC(\emptyset) \quad \Rightarrow \quad h^e(\alpha) \in FC_2(\emptyset), \qquad \text{for each } e\colon At \to S_2$$

and suppose that $\alpha \notin FC_2(X)$. By A.3, we have $FC_2(X \cup \{\sim \alpha\}) \neq S_2$, hence we conclude from (iii) that there exists a valuation $v\colon At \to \{0,1\}$ such that $h^v(X \cup \{\sim \alpha\}) \subseteq \{1\}$. Let $e\colon At \to S_2$ be defined as

$$e(\gamma) = \begin{cases} \gamma \to \gamma & \text{if } v(\gamma) = 1 \\ \sim(\gamma \to \gamma) & \text{if } v(\gamma) = 0. \end{cases}$$

It is easy to observe that for each $\phi \in S_2$,

$$\begin{aligned} h^v(\phi) = 1 \quad &\Rightarrow \quad h^e(\phi) \in FC_2(\emptyset), \\ h^v(\phi) = 0 \quad &\Rightarrow \quad \sim h^e(\phi) \in FC_2(\emptyset). \end{aligned}$$

The easy proof is by induction on the length of the formula ϕ. Then we conclude that $h^e(X \cup \{\sim \alpha\}) \subseteq FC_2(\emptyset)$ which is incompatible with $(\star)$.

(iv)$\Rightarrow$(v): Assume that $\alpha \notin FC_2(X)$. From (iv) we get $h^e(X) \subseteq FC_2(\emptyset)$ for some $e\colon At \to S_2$. Hence, for each $v\colon At \to \{0,1\}$, we have $h^v\left(h^e(X)\right) \subseteq \{1\}$. Let us put $Y_v = \{\phi \in S_2 : h^v\left(h^e(\phi)\right) = 1\}$. It is evident (see Lemma A.1) that the set

Y_v fulfills the consequent of the implication (v).

(v)$\Rightarrow$(vi): Let us consider $\mathscr{L}_2(X)$ (for $FC_2(X) \neq S_2$) and let M be the family of those FC_2-maximal sets Y which contain X. We recall that $Y = FC_2(Y) \neq S_2$ is a FC_2-maximal iff $\phi \in Y$ or $\sim \phi \in Y$ for each formula $\phi \in S_2$. We are going to show that the algebra $\mathscr{L}_2(X)$ is isomorphic with some field of sets on the just defined set M. This isomorphism is defined as follows: $i([\phi]) = \{Y \in M : \phi \in Y\}$.

(a) $i([\phi] \cap [\psi]) = i([\phi]) \cap i([\psi])$.

The above is quite obvious. Let us prove

(b) $i([-\phi]) = -i([\phi])$.

Suppose that $Y \in i(-[\phi]) = i([\sim \phi])$. Hence $\sim \phi \in Y$ and therefore $Y \notin i([\phi])$ since $Y = FC_2(Y) \neq S_2$. Assume next that $Y \in -i([\phi])$, i.e., $\phi \notin Y$. Hence $\sim \phi \in Y$ because Y is FC_2-maximal and therefore $Y \in i(| \sim \phi|) = i(-[\phi])$.

By (a) and (b) it can be concluded that $i \colon S_2 / \sim_X \to 2^M$ is a homomorphism. Moreover the mapping i is one-to-one:

(c) $[\alpha] \neq [\beta] \Rightarrow i([\alpha]) \neq i([\beta])$.

Indeed, let $[\alpha] \neq [\beta]$. We have then $(\alpha \equiv \beta) \notin FC_2(X)$, e.g., $\beta \notin FC_2(X \cup \{\alpha\})$. From this we obtain (see Lemma A.3) $FC_2(X \cup \{\alpha, \sim \beta\}) \neq S_2$, hence by (v) there exists a maximal set Y such that $X \cup \{\alpha, \sim \beta\} \subseteq Y$, so $Y \in i([\alpha]) \setminus i([\beta])$. Since i is a one-to-one homomorphism, the algebra $\mathscr{L}_2(X)$ is isomorphic with some subalgebra of 2^M, that is with some field of sets.

(vi)$\Rightarrow$(i): This implication results from Lemma A.2. $\square$

The Axiom of Choice AC has not been used in the above proof though the axiom is necessary for proving each of the statements (i)–(vi) separately. Note that (ii)–(vi) can be proved effectively (i.e., without using AC) if we assume that the set of variables $At \subseteq S_2$ is countable. The above list of theorems can be easily extended by adding, for instance, the theorem on the existence of an ultrafilter in any Boolean algebra; the so-called BPI theorem. Our aim, however, is to present only logical equivalents of BPI. The equivalence of (i), (iii), (v) (vi) was proved by Łoś and Henkin (see [62] 1954, [63] 1957, [40] 1953). Let us rewrite Metatheorem A.4 replacing formal statements (i)–(vi) with their informal names.

Metatheorem A.4. *The following theorems are effectively equivalent:*

(i) *Stone's representation theorem for Boolean algebras.*

(ii) *Strong adequacy of the two-element Boolean algebra (or matrix $\mathfrak{M}_2$) for the classical propositional logic.*

(iii) *Gödel–Malcev's propositional theorem.*

(iv) *Structural completeness theorem for the classical propositional logic.*

(v) *Lindenbaum–Łoś's maximalization theorem.*

(vi) *Łoś's theorem on the representation of Lindenbaum–Tarski algebras.*

Appendix B
A proof system for the classical logic

We define an operation on inferential rules, called *Hilbert's operation* (or *H-operation*, in short). Given two inferential rules, say r and r', one can produce by use of it another rule denoted by $r(r')$. The operation was defined by Pogorzelski [81], 1975, but some basic ideas came from Hilbert's proof of 'Prämissentheorem', see [42], 1939. We use this operation to minimize the axiom basis for the classical logic.

H-operation

Let us recall that S denotes the set of formulas and $\mathscr{R}$ stands for the set of inferential rules in a given propositional language. In the present chapter we confine ourselves to rules with finite, non-empty sets of premisses. Let $\mathscr{R}^+$ be the set of standard rules (i.e., rules with basic sequents) determined by sequents of the form $\langle \{p_0\}, \Phi(p_0, p_1, \ldots p_k) \rangle$. Note that $r_1 \in \mathscr{R}^+$ where

$$r_1 : \frac{\phi}{\psi \to \phi}$$

is the rule determined by the sequent $\langle \{p_0\}, p_1 \to p_0 \rangle$ in the standard propositional language S_2. It should also be clear that the modus ponens rule

$$r_0 : \frac{\phi, \ \phi \to \psi}{\psi}$$

does not belong to $\mathscr{R}^+$.

Suppose that $r \in \mathscr{R}^+$ and $r' \in \mathscr{R}$. Let

$$r : \frac{\alpha}{\Phi(\alpha, \beta_1, \ldots, \beta_k)} \quad \text{for all} \quad \alpha, \beta_1, \ldots, \beta_k \in S.$$

Now, we can define a new rule $r(r')$. This operation can be viewed as a passage

$$\text{from} \quad r' : \frac{\alpha_1, \ldots, \alpha_n}{\alpha} \quad \text{to} \quad r(r') : \frac{\Phi(\alpha_1, \beta_1, \ldots, \beta_k), \ldots, \Phi(\alpha_n, \beta_1, \ldots, \beta_k)}{\Phi(\alpha, \beta_1, \ldots, \beta_k)}.$$

In other words, the rule $r(r')$ consists of all sequents of the form

$$\langle \{\Phi(\alpha_1, \beta_1, \ldots, \beta_k), \ldots, \Phi(\alpha_n, \beta_1, \ldots, \beta_k)\}, \Phi(\alpha, \beta_1, \ldots, \beta_k) \rangle$$

where $\langle \{\alpha_1, \ldots, \alpha_n\}, \alpha \rangle \in r'$ and $\beta_1, \ldots, \beta_k \in S$. The rule $r \in \mathscr{R}^+$ may have many basic sequents — each alphabetical variant of a basic sequent is also a basic sequent — but the above definition does not depend on the choice of the alphabetical variant of $\Phi(p_0, p_1, \ldots, p_k)$. Notice that, if r' is standard, then $r(r')$ is standard as well and its basic sequent is

$$\langle \{\Phi(\gamma_1, p_1, \ldots p_k), \ldots, \Phi(\gamma_n, p_1, \ldots, p_k)\}, \Phi(\gamma, p_1, \ldots, p_k) \rangle$$

where $\langle \{\gamma_1, \ldots, \gamma_n\}, \gamma \rangle$ is a basic sequent for r' and the variables $p_1, \ldots, p_k$ do not occur in the formulas γ_i. Let us give some examples. One easily checks that

$$r_1(r_0) : \frac{\beta \rightarrow \phi, \ \beta \rightarrow (\phi \rightarrow \psi)}{\beta \rightarrow \psi}.$$

Next, note that if

$$r : \frac{\alpha}{(\delta \rightarrow \sim \alpha) \rightarrow \delta} \quad \text{and} \quad r' : \frac{\beta \rightarrow \sim \gamma}{\gamma \rightarrow \sim \beta},$$

then

$$r(r') : \frac{(\delta \rightarrow \sim (\beta \rightarrow \sim \gamma)) \rightarrow \delta}{(\delta \rightarrow \sim (\gamma \rightarrow \sim \beta)) \rightarrow \delta}.$$

The basic property of the classical propositional logic is its closure under H-operation:

Theorem B.1. *For every* $r \in \mathscr{R}^+$ *and* $r' \in \mathscr{R}$,

$$r, r' \in \operatorname{Der}(R_0, Sb(A_2)) \quad \Rightarrow \quad r(r') \in \operatorname{Der}(R_0, Sb(A_2)).$$

Proof. Suppose that $r, r' \in \operatorname{Der}(R_0, Sb(A_2))$ and let $\langle \{p_0\}, \Phi(p_0, p_1, \ldots, p_k) \rangle$ be a basic sequent for r. If $r(r') \notin \operatorname{Der}(R_0, Sb(A_2))$, we would have

$$\Phi(\alpha, \beta_1, \ldots \beta_k)) \notin Cn(R_0, Sb(A_2) \cup \{\Phi(\alpha_1, \beta_1, \ldots, \beta_k), \ldots, \Phi(\alpha_n, \beta_1, \ldots, \beta_k)\})$$

for some $\langle \{\alpha_1, \ldots, \alpha_n\}, \ \alpha \rangle \in r'$ and some $\beta_1, \ldots, \beta_k$.

Let $\mathfrak{M}_2$ be the two-element matrix strongly adequate for the classical propositional logic. Let $0, 1$ denote, as usual, the elements of the matrix. By the completeness theorem, we would get a valuation $v : At \rightarrow \{0, 1\}$ such that

$$h^v(\Phi(\alpha, \beta_1, \ldots, \beta_k)) = 0 \quad \text{and} \quad h^v(\Phi(\alpha_i, \beta_1, \ldots, \beta_k)) = 1 \quad \text{for each} \ i \leqslant n.$$

Since $\langle\{\alpha\}, \Phi(\alpha, \beta_1, \ldots, \beta_k)\rangle \in r$ and $r \in \mathrm{Der}\big(R_0, Sb(A_2)\big)$, we have

$$\Phi(\alpha, \beta_1, \ldots, \beta_k) \in Cn\big(R_0, Sb(A_2) \cup \{\alpha\}\big).$$

But $h^v\big(\Phi(\alpha, \beta_1, \ldots, \beta_k)\big) = 0$, so $h^v(\alpha) = 0$. The rule r' is derivable in the classical logic, hence $h^v(\alpha_i) = 0$ for some $i \leqslant n$. Thus, there is an $i \leqslant n$ such that $h^v\big(\Phi(\alpha, \beta_1, \ldots, \beta_k)\big) = h^v\big(\Phi(\alpha_i, \beta_1, \ldots, \beta_k)\big)$ which is impossible. □

Suppose that a propositional system $\langle R, A \rangle$ is given. We would like to extend the usual derivation on the basis of the system, that is the consequence Cn_{RA}, with H-operation. It means that we would like to use the operation in derivations (formal proofs), in addition to standard applications of the rules R and axioms A. It can be achieved formally in several ways. One of them is to define inductively an operation $C_{RA}: 2^S \to 2^S$ and an auxiliary set of rules $D(R, A)$. We will write $C(X)$ instead of $C_{RA}(X)$, and $D(C)$ instead of $D(R, A)$. We hope this does not lead to confusion. Let

(i)	$\alpha \in C^0(X)$	$\Leftrightarrow$ $\alpha \in A \cup X$,
(ii)	$r \in D^0(C)$	$\Leftrightarrow$ $r \in R$,
(iii)	$\alpha \in C^{n+1}(X)$	$\Leftrightarrow$ $\alpha \in C^n(X) \vee \langle\Pi, \alpha\rangle \in r$ for some $\Pi \subseteq C^n(X)$ and some $r \in D^n(C)$,
(iv)	$r \in D^{n+1}(C)$	$\Leftrightarrow$ $r \in D^n(C) \vee \alpha \in C^n(\Pi)$ for all $\langle\Pi, \alpha\rangle \in r \vee$ $\vee \big(r' \in \mathscr{R}^+ \wedge r = r'(r'')\big)$ for some $r', r'' \in D^n(C)$,
(v)	$\alpha \in C(X)$	$\Leftrightarrow$ $\alpha \in C^n(X)$ for some n,
(vi)	$r \in D(C)$	$\Leftrightarrow$ $r \in D^n(C)$ for some n.

The above definition, although complicated, is quite standard, except for a fragment of (iv) where H-operation appears. Without problems one should prove the following corollary which provides us with another definition of C_{RA}:

Corollary B.2. *C_{RA} is the least consequence operation which is closed under H-operation and $Cn_{RA} \leqslant C_{RA}$. Moreover,* $\mathrm{Der}(C_{RA}) = D(C_{RA})$.

The two-valued $\{\to\}$ logic

Let us consider a propositional system (R_0, A_0), in the pure implicational language $\{\to\}$, where $R_0 = \{r_0\}$ and A_0 consist of one axiom schema:

(c_1) $\qquad\qquad\qquad\qquad\qquad \alpha \to (\beta \to \alpha).$

The C-like consequence operation (see the previous section) defined by the system is denoted by C_0, that is $C_0(X) = C_{R_0, A_0}(X)$ for each X. Let us prove that this consequence operation coincides with the consequence operation determined — in the standard manner — by the classical logic $\{\to\}$:

Theorem B.3. $C_0(X) = Cn\big(R_0, Sb(A_2^\to) \cup X\big),$ *for every set $X \subseteq S^\to$.*

Proof. The inclusion ($\subseteq$) is obvious as (R_0, A_0) is a subsystem of the classical logic and it has been proved that the classical logic is closed under H-operation. For the other, it is sufficient to derive, using the definition of the operation C_0, all axioms of the pure implicational fragment of classical logic. As axioms we can take, for instance, the formulas (c_1)–(c_4) where:

(c_2) $$(\gamma \to (\gamma \to \beta)) \to (\gamma \to \beta),$$

(c_3) $$(\delta \to \alpha) \to ((\alpha \to \gamma) \to (\delta \to \gamma)),$$

(c_4) $$((\alpha \to \beta) \to \beta) \to ((\beta \to \alpha) \to \alpha).$$

(c_2): We derive — in the standard manner — the rule r_1. Then, we also get $r_1(r_0)$, that is we get $r_1(r_0) \in D(C_0)$. Let $r_2 = r_1(r_0)$. We have

$$r_2 : \frac{\gamma \to \alpha \, , \, \gamma \to (\alpha \to \beta)}{\gamma \to \beta}.$$

If we take γ for β and $\alpha \to \gamma$ for α, we receive $\gamma \to \gamma \in C_0(\emptyset)$. Then, by r_2,

$$\frac{\gamma \to (\gamma \to \beta)}{\gamma \to \beta}$$

and, next, we receive c_2 using r_1 and $(\gamma \to (\gamma \to \alpha)) \to (\gamma \to (\gamma \to \alpha))$.

(c_3): Notice that the following is a subrule of r_2:

$$\frac{(\alpha \to \gamma) \to \alpha \, , \, (\alpha \to \gamma) \to (\alpha \to \gamma)}{(\alpha \to \gamma) \to \gamma}$$

and hence we also get the rule

$$\frac{\alpha}{(\alpha \to \gamma) \to \gamma}$$

and the rule

$$r_3 : \frac{\delta \to \alpha}{\delta \to ((\alpha \to \gamma) \to \gamma)}.$$

On the other hand, using $r_1(r_2)$, we get

$$\frac{\beta \to (\alpha \to (\beta \to \gamma)) \, , \, \beta \to (\alpha \to \beta)}{(\beta \to (\alpha \to \gamma)}$$

which gives us the rule

$$r_4 : \frac{\alpha \to (\beta \to \gamma)}{\beta \to (\alpha \to \gamma)}.$$

Using r_3 and r_4 we receive — in the standard way —

$$\frac{\delta \to \alpha}{(\alpha \to \gamma) \to (\delta \to \gamma)}$$

which gives us (c_3) on the basis of r_1.

(c_4): Using r_1 we show in the standard manner

$$r_5 : \frac{\alpha}{((\alpha \to \gamma) \to \gamma) \to ((\gamma \to \alpha) \to \alpha)}.$$

Hence, as $r_5(r_0)$, we receive the rule

$$((\alpha \to \gamma) \to \gamma) \to ((\gamma \to \alpha) \to \alpha)$$

$$\frac{(((\alpha \to \beta) \to \gamma) \to \gamma) \to ((\gamma \to (\alpha \to \beta)) \to (\alpha \to \beta))}{((\beta \to \gamma) \to \gamma) \to ((\gamma \to \beta) \to \beta)}.$$

Thus, we obtain (c_4) if we identify α with γ.   □

It also follows from the above theorem that $D(C_0) = \mathrm{Der}(R_0, Sb(A_2^{\to}))$. Let us note that one could prove a similar result for a weaker (as concerns the traditional derivation) propositional system consisting of the two rules $\{r_0, r_1\}$ and one axiom schema $\alpha \to \alpha$.

The two-valued $\{\to, \sim\}$ logic

Let us consider the language $\{\to, \sim\}$ and the system (R_0, A_1), where $R_0 = \{r_0\}$ and A_1 consists of (c_1) and

(nc_1) $\qquad\qquad\qquad\qquad\qquad \alpha \to (\sim \alpha \to \beta),$

(nc_2) $\qquad\qquad\qquad\qquad\qquad \sim\sim (\alpha \to \alpha).$

Let C_1 be the C-like consequence operation determined by the above system. Let us prove that C_1 coincides with the consequence operation determined by the classical logic in $\{\to, \sim\}$:

Theorem B.4. $C_1(X) = Cn(R_0, Sb(A_2) \cup X)$, for every set $X \subseteq S^{\to,\sim}$.

Proof. Note that our system is an extension of the system (R_0, A_0) considered in the previous chapter. Therefore $C_1(\emptyset)$ contains all $\{\to, \sim\}$ instances of all implicational tautologies. Thus, to prove our theorem, it suffices to derive in our system, for instance, $(\sim \alpha \to\sim \beta) \to (\beta \to \alpha)$.

Let

$$r_6 : \frac{\alpha}{\sim \alpha \to \gamma}, \qquad\qquad r_7 : \frac{\sim \alpha \to \gamma, \ \sim (\alpha \to \beta) \to \gamma}{\sim \beta \to \gamma}.$$

Both rules are derivable for C_1, by (nc_1) and the fact that $r_7 = r_6(r_0)$. We substitute $\sim \alpha$ for γ in r_7 and we get

$$\frac{\sim (\alpha \to \beta) \to \sim \alpha}{\sim \beta \to \sim \alpha}.$$

Then we apply r_6 to receive, from the above, the rule

$$\frac{\alpha \to \beta}{\sim \beta \to \sim \alpha}$$

and hence $(\alpha \to \beta) \to (\sim \beta \to \sim \alpha)$ as well as $(\sim \alpha \to \sim \beta) \to (\sim\sim \beta \to \sim\sim \alpha)$. Then, by (nc_1) and (nc_2) (take $\alpha \to \alpha$ for β in the above formula), we also get $\alpha \to \sim\sim \alpha$ which gives us

$(*)$ $\qquad\qquad\qquad\qquad (\alpha \to \sim \beta) \to (\beta \to \sim \alpha).$

The following formula is an (instance of an) implicational tautology and hence it is derivable in our system:

$$((\alpha \to (\sim \alpha \to \sim (\alpha \to \sim \alpha)))) \to ((\alpha \to \sim \alpha) \to (\alpha \to \sim (\alpha \to \sim \alpha))).$$

Then, by (nc_1) and r_0, we get

$$(\alpha \to \sim \alpha) \to (\alpha \to \sim (\alpha \to \sim \alpha))$$

which gives us, by $(*)$,

$$(\alpha \to \sim \alpha) \to ((\alpha \to \sim \alpha) \to \sim \alpha)$$

Thus, we also get $(\alpha \to \sim \alpha) \to \sim \alpha$ which is equivalent (on the basis of an implicational tautology) to $(\sim \alpha \to \alpha) \to \alpha$. Then we get $\sim\sim \alpha \to \alpha$ by (nc_1) and hence, using $(*)$ again, we get $(\sim \alpha \to \sim \beta) \to (\beta \to \alpha)$. $\qquad\square$

One could replace nc_2 with any classical tautology of the form $\sim \alpha$ but we can also show that (c_1) and (nc_1), together with r_0, are not sufficient to axiomatize Cn_2 — the consequence operation determined by the classical logic. Indeed, let us consider a two-valued logic, say, Cn'_2 in which $\sim$ is interpreted as the falsum operator. The logic Cn'_2 is H-closed by Theorem B.1. Their intersection $Cn_2 \cap Cn'_2$ is H-closed, as well. Since (c_1), (nc_1) and r_0 are valid for both logics, we conclude that the axiom schema (nc_2) cannot be removed from A_1.

Let us consider a three-element matrix $\mathfrak{M}_D$, on the set $\{1, 2, 3\}$, with 3 as the only designated element, and the operators interpreted as:

$\to$	1	2	3
1	3	3	3
2	3	3	3
3	1	2	3

$\sim$	
1	3
2	2
3	1

All $\{\rightarrow\}$-tautologies, as well as the formulas A_1 and the rule r_0 are valid in this matrix. On the other hand, none of the following formulas is valid there:

$$(\alpha \rightarrow \beta) \rightarrow (\sim \beta \rightarrow \sim \alpha)$$

$$(\alpha \rightarrow \sim \beta) \rightarrow (\beta \rightarrow \sim \alpha)$$

$$(\sim \alpha \rightarrow \sim \beta) \rightarrow (\beta \rightarrow \alpha)$$

$$(\sim \alpha \rightarrow \beta) \rightarrow (\sim \beta \rightarrow \alpha)$$

$$(\alpha \rightarrow \beta) \rightarrow ((\alpha \rightarrow \sim \beta) \rightarrow \sim \alpha)$$

$$(\alpha \rightarrow \beta) \rightarrow ((\sim \alpha \rightarrow \beta) \rightarrow \beta)$$

$$(\sim \alpha \rightarrow \alpha) \rightarrow \alpha$$

$$(\alpha \rightarrow \sim \alpha) \rightarrow \sim \alpha$$

This shows that for the traditional notion of derivability, that is for Cn_{RA}, the set A_1 is relatively weak. In particular, the above formulas cannot be deduced from (nc_1) and (nc_2) even on the basis of the pure implicative fragment of classical logic.

H-completeness

A propositional system $\langle R, A \rangle$ is said to be *H-complete* if its derivable rules are closed under the H-operation, i.e.,

$$r, r' \in \mathrm{Der}(R, A) \quad \Rightarrow \quad r(r') \in \mathrm{Der}(R, A)$$

for every $r \in \mathscr{R}^+$ and $r' \in \mathscr{R}$.

The above property is rare, see Theorem B.1. One easily notices that the least propositional logic (with the empty set of rules and axioms) does possess this property. But it is not easy to identify other systems of this kind. In the proof of Theorem B.1 we have not used any specific property of classical logic (except for completeness). Thus, one could repeat the above argument for each propositional logic determined by a matrix with at most one undesignated element. It means, in particular, that each two-valued logic, independently of the language considered, is closed under H-operation. Clearly, any intersection of logics which are closed under H-operation is also closed under the operation. It gives us a number of logics which fulfill the above property. But it is not easy to give a natural (that means considered in literature) non-classical system which is included in the family. In view of Theorem B.2, there is no H-complete subsystem of the classical logic containing $\langle R_0, A_0 \rangle$.

Theorem B.5. *An invariant propositional logic $\langle R, A \rangle$ (we write Cn for Cn_{RA}) is H-complete if it fulfills the following conditions:*

(i) *If* $\Phi \in Cn(\{p_0\})$, *then* $Cn(\{\Phi\}) = Cn(\{p_0 + \Psi\})$ *for some* Ψ *which does not contain the variable* p_0.

(ii) *If* $\alpha_0 \in Cn(\{\alpha_1, \ldots, \alpha_n\})$ *then* $(\alpha_0 + \Psi) \in Cn(\{\alpha_1 + \Psi, \ldots, \alpha_n + \Psi\})$, *for each of the formulas* $\alpha_0, \ldots, \alpha_n, \Psi$.

Proof. Suppose that $r' \in \mathrm{Der}(Cn)$ and let $\langle \{\alpha_1, \ldots, \alpha_n\}, \alpha_0 \rangle \in r'$. Then we have $\alpha_0 \in Cn(\{\alpha_1, \ldots, \alpha_n\})$. Assuming that $r \in \mathrm{Der}(Cn)$ and $\langle \{p_0\}, \Phi \rangle$ is a basic sequence for r, we conclude by (i) that $\Phi[p_0/\alpha_i]$ is equivalent to $\alpha_i + \Psi$, for each i, and hence $\Phi[p_0/\alpha_0] \in Cn(\{\Phi[p_0/\alpha_1] \ldots, \Phi[p_0/\alpha_n]\})$ by (2).   □

In contrast to Theorem B.1, the above result does make a use of certain properties of classical operators. The property (ii) is the usual characterization of disjunction and it is fulfilled by many logical systems. The property (i) is not so usual. Using the above theorem one can show, for instance, that the classical predicate logic is H-complete.

It should also be noted that H-completeness differs from all the variants of completeness which were considered in the previous chapters. In particular, it does not coincide with the notion of Γ-maximality (in the family of Γ-invariant systems). We have already produced above H-complete systems which are not maximal in the family of all consistent invariant logics. On the other hand side, one can also give an example of a maximal invariant logic which is not H-complete. Such systems $\langle R, A \rangle$ are *H-inconsistent* which means that they generate the inconsistent consequence operation C_{RA} (even though the standard consequence operations Cn_{RA} might be consistent).

Bibliography

[1] Asser G., *Einführung in die Mathematische Logik*, Teil I *Aussagenkalkül*, B.G. Teubner Verlagsgesellschaft, Leipzig 1959 (4th edition 1972).

[2] Bernert J., *A note on existing of matrices strongly adequate for some propositional logics*; Reports on Mathematical Logic 14 (1982), pp. 3–7.

[3] Biela A., *Note on the structural incompleteness of some modal propositional calculi*; Reports on Mathematical Logic 4 (1975), pp. 3–6.

[4] Biela A., *On the so–called Tarski's property in the theory of Lindenbaum oversystems*, part I; Reports on Mathematical Logic 7 (1976), pp. 3–20; part II, ibid. 11 (1981), pp. 13–48.

[5] Biela A., Dziobiak W., *On two properties of structural completeness*; Reports on Mathematical Logic 16 (1983), pp. 51-54.

[6] Biela A., Pogorzelski W.A., *The power of the class of Lindenbaum–Asser extensions of consistent set of formulas*; Reports on Mathematical Logic 2 (1974), pp. 5–8.

[7] Blok W.J., *On the degree of incompleteness of modal logics*; Bulletin of the Section of Logic 7 (1978), pp. 167–175.

[8] Blok W.J., Pigozzi D., *Algebraizable Logics*; Memoirs of the AMS 77, nr 396(1989).

[9] Bloom S.L., *A representation theorem for standard consequence operations*; Studia Logica 34 (1975), pp. 235–238.

[10] Chang C.C., *A new proof of the completeness of the Łukasiewicz axioms*; Transactions of the American Mathematical Society 93 (1959), pp. 74–80.

[11] Cignoli R., Mundici D., *An elementary proof of the Chang's completeness theorem for the infinite–valued calculus of Łukasiewicz*; Studia Logica 58 (1997), pp. 79–97.

[12] Citkin A.I., *On admissible rules in the intuitionistic propositional logic*; Matematiceskij Sbornik 102 (1977), pp. 314–323.

[13] Czelakowski J., *Model-theoretic methods in methodology of propositional calculi*; Polish Academy of Sciences, Institute of Philosophy and Sociology, Warsaw 1980.

[14] Czelakowski J., *Logical matrices and the amalgamation property*; Studia Logica 41 (1982), pp. 329–342.

[15] Czelakowski J., *Matrices, primitive satisfaction and finitely based logics*; Studia Logica 42 (1983), pp. 89–10.

[16] Dummett M., *A propositional calculus with denumerable matrix*; The Journal of Symbolic Logic 24 (1959), pp. 97–106.

[17] Dunn J.M., Meyer R.K., *Algebraic completeness results for Dummett's LC and its extensions*; Zeitschrift für Mathematische Logik und Grundlagen der Mathematik 7 (1971), pp. 225–230.

[18] Dzik W., Wroński A., *Structural completeness of Gödel's and Dummett's propositional calculi*; Studia Logica 32 (1973), pp. 69–73.

[19] Dziobiak W., *An example of strongly finite consequence operation with $2^{\aleph_0}$ standard strengthenings*; Studia Logica 39 (1980), pp. 375–379.

[20] Dziobiak W., *The lattice of strengthenings of strongly finite consequence operation*; Studia Logica 40 (1981), pp. 177–194.

[21] Dziobiak W., *The degree of maximality of the intuitionistic propositional logic and some of its fragments*; Studia Logica 40 (1981), pp. 195–198.

[22] Dziobiak W., *A finite matrix whose set of tautologies is not finitely axiomatizable*; Reports on Mathematical Logic 25 (1991), pp. 105–112.

[23] Feys R., *Modal Logics*; Gauthier-Villars, Paris 1965.

[24] Frink O., *Ideals in partially ordered sets*; American Mathematical Monthly, 61 (1954), pp. 223–234.

[25] Ghilardi S., *Unification in intuitionistic logic*; The Journal of Symbolic Logic, 64 (1999), pp. 859–80.

[26] Gödel K., *Eine Interpretation des intuitionistischen Aussagenkalküls*; Ergebnisse eines Mathematischen Kolloquiums 4 (1933), pp. 39–40.

[27] Grabowski A., *Lattice of substitutions*; Reports on Mathematical Logic, 33 (1999), pp. 99–109.

[28] Grätzer G., *General lattice theory*; Birkhäuser Verlag, Basel und Stuttgart 1978.

[29] Grygiel J., *Absolutely independent axiomatizations for countable sets in classical logic*, Studia Logica 48(1989), pp. 77–84.

[30] Grygiel J., *Absolutely independent sets of generators of filters in Boolean algebras*; Reports on Mathematical Logic 24(1991), pp. 25–35.

[31] Grygiel J., *Freely generated filters in free Boolean algebras*; Studia Logica 54(1994), pp. 139–147.

[32] Grygiel J., *Scarce decomposition of finite distributive lattices*; Contributions to General Algebra 13(2001), pp. 149–158.

[33] Grygiel J., *Application of the Cantor-Bendixon construction to the problem of freely generated filters*; Reports on Mathematical Logic, 38(2004), pp. 49–59.

[34] Grygiel J., *Distributive lattices with a given skeleton*; Discussiones Mathematicae, General Algebra and Applications, 24(2004), pp. 75–94.

[35] Grygiel J., *Some numerical characterization of finite distributive lattices*; Bulletin of the Section of Logic, 33/3(2004), pp. 1–7.

[36] Grygiel J., *The Concept of Gluing for Lattices*; Pedagogical Academy in Częstochowa 2004.

[37] Grzegorczyk A., *An approach to logical calculus*; Studia Logica 30 (1972), pp. 33–43.

[38] Hao Wang, *Note on rules of inference*; Zeitschrift für Mathematische Logik und Grundlagen der Mathematik 11 (1965), pp. 193–196.

[39] Hawranek J., Zygmunt J., *On the degree of complexity of sentential logic. A couple of examples*; Studia Logica 40 (1981), pp. 141–154. Part two of the paper; Studia Logica 43(1984), pp.405–413.

[40] Henkin L., *Some interconnections between modern algebra and mathematical logic*; Translations of American Mathematical Society 74 (1953), pp. 410–427.

[41] Herrmann Ch., *S-verklebte Summen von Verbänden*; Mathematische Zeitschrift 130 (1973), pp. 255–274.

[42] Hilbert D., Bernays P., *Grundlagen der Mathematik*, Band II; Springer Verlag, Berlin 1939.

[43] Iemhoff R., *On the admissible rules in intuitionistic propositional logic*; The Journal of Symbolic Logic 66 (2001), pp. 281–284.

[44] Jaśkowski S., *Recherches sur le système de la logique intuitioniste*; Actes du Congrès International de Philosophie Scientifique, VI Philosophie des Mathématiques, Actualités scientifiques et industrielles 393 (1936), pp. 58–61.

[45] Kolany A., *Satisfiability on Hypergraphs*; Studia Logica 52(3) (1993), pp. 393–404.

[46] Kolany A., *Hypergraphs and the Intuitionistic Propositional Calculus* ; Reports on Mathematical Logic, 27 (1993), pp. 55–66.

[47] Kolany A., *A General Method of Solving SmullyanŠs Puzzles* ; Logic and Logical Philosophy, 4 (1996), pp. 97–103.

[48] Kolany A., *Consequence Operations Based on Hypergraph Satisfiability*, Studia Logica, 58(2) (1997), pp. 261–272.

[49] Kolany A., *Grabowski Lattices are Generated by Graphs*; Reports on Mathematical Logic, 36 (2002), pp. 63–69.

[50] Kolany A., *Lattices of non-Locally Finite Hypergraphs are not Heyting* ; Bulletin of the Section of Logic, 35/2-3 (2006), pp. 105–110

[51] Kostrzycka Z., *On the density of implicational parts of intuitionistic and classical logics*; Journal of Applied Non-Classical Logics 13 (2003), pp. 295–325.

[52] Kostrzycka Z., *On the density if truth in Dummet's logic*; Bulletin of the Section of Logic 32 (2003), pp. 43–55.

[53] Kostrzycka Z., *On the density of truth in Grzegorczyk's modal logic*; Bulletin of the Section of Logic, 33 (2004), pp. 107–120.

[54] Kostrzycka Z., *The density if truth in monadic fragment of some intermediate logics*; Journal of Logic, Language and Information 16 (2007), pp. 283–302.

[55] Kostrzycka Z., *On the existence of a continuum of logics in NEXT(KTB)*; Bulletin of the Section of Logic 36/1 (2007), pp. 1–7.

[56] Kostrzycka Z., Zaionc M., *Statistics of intuitionistic versus classical logic*; Studia Logica 76 (2004), pp. 307–328.

[57] Kotas J., Wojtylak P., *Finite distributive lattices as sums of Boolean algebras*; , Reports on Mathematical Logic 29 (1995), pp. 35–40.

[58] Latocha P., *The problem of structural completeness of intuitionistic propositional logic*; Reports on Mathematical Logic 16 (1983), pp.17-22.

[59] Lemmon E.J., Meredith C.A., Meredith D., Prior A.N., Thomas I., *Calculi of pure strict implication*: Philosophical Department of Canterbury University, Christchurch, New Zealand 1957 (mimeographed); in: Philosophical Logic, edited by J.W.Davis et al., D. Reidel Publishing Company 1969, pp. 215–250.

[60] Lewis C.I., Langford C.H., *Symbolic Logic*; New York 1932.

[61] Łoś J., *An algebraic proof of completeness for the two–valued propositional calculus*; Colloquium Mathematicum 2 (1951), pp. 236–240.

[62] Łoś J., *Sur le théorème de Gödel pour les théories indénombrales*; Bulletin de l'Académie Polonaise des Sciences, Cl. III, 2 (1954), pp. 319–320.

[63] Łoś J., *Remarks on Henkin's paper: Boolean representation through propositional calculus*; Fundamenta Mathematicae 44 (1957), pp. 82–83.

[64] Łoś J., Suszko R., *Remarks on sentential logics*; Indagationes Mathematicae 20 (1958), pp. 177–183.

[65] Łukasiewicz J., *Elements of Mathematical Logic* (in Polish); Warszawa 1929.

[66] Łukasiewicz J., Tarski A., *Untersuchungen über den Aussagenkalkül*; Comptes rendus des séances de la Société des Sciences et des Lettres de Varsovie, Cl III, 23 (1930), pp. 30–50.

[67] Maduch M., *The consequence operation determined by Cartesian product of matrices* (in Polish); Zeszyty Naukowe WSP Opole, Matematyka 13 (1973), pp. 159–162.

[68] Makinson D., *A characterization of structural completeness*; Reports on Mathematical Logic 6 (1976), pp. 99–102.

[69] Maksimova L.L., *The principle of separating variable in sentential logics* (in Russian); Algebra i Logika 15 (1976), pp. 168–184.

[70] Malinowski G., *Degrees of maximality of Łukasiewicz–like sentential calculi*; Studia Logica 34 (1997), pp. 213–228.

[71] Minc G.E., *Derivability of admissible rules*; Zapiski Nauchnego Seminaria Leningrad Otdel. Math. Ins. Steklov (LOMI) 32 (1972), pp.85–89; English translation in Journal of Soviet Mathematics 6 (1976).

[72] Pałasińska K., *Three–element nonfinitely axiomatizable matrices*; Studia Logica 53 (1994), pp. 361–372.

[73] Pogorzelski W.A., *The deduction theorem for Łukasiewicz many–valued propositional calculi*; Studia Logica 15 (1964), pp. 7–23.

[74] Pogorzelski W.A., *On the scope of the classical deduction theorem*; Journal of Symbolic Logic 33, 1968.

[75] Pogorzelski W.A., *Some remarks on the notion of completeness of the propositional calculus* (in Polish); Studia Logica 23 (1968), pp. 43–50.

[76] Pogorzelski W.A., *Note on the power of the so–called relative Lindenbaum's extensions* (in Polish); Prace z Logiki, Zeszyty naukowe UJ 47 (1969), pp. 25–29.

[77] Pogorzelski W.A., *The two–valued propositional calculus and the deduction theorem*; Acta Universitatis Vratislaviensis 101 (1969), pp. 13–18.

[78] Pogorzelski W.A., *The classical propositional calculus* (in Polish); PWN Warszawa 1969 (3rd edition 1975).

[79] Pogorzelski W.A., *Structural completeness of the propositional calculus*; Bulletin de l'Académie Polonaise des Sciences, série des Sciences Mathématiques, Astronomiques et Physiques 19 (1971), pp. 349–351.

[80] Pogorzelski W.A., *Concerning the notion of completeness of invariant sentential calculi*; Studia Logica 33 (1974), pp. 69–72.

[81] Pogorzelski W.A., *On Hilbert's operation on logical rules* I; Reports on Mathematical Logic 12 (1981), pp. 35–50.

[82] Pogorzelski W.A., *On Hilbert's operation on logical rules* II; Reports on Mathematical Logic 12 (1984), pp. 3–11.

[83] Pogorzelski W.A., *Notions and Theorems of Elementary Formal Logic*; Published by Warsaw University-Białystok Branch, Białystok 1994.

[84] Pogorzelski W.A., Prucnal T., *Some remarks on the notion of completeness of the propositional calculus*; Reports on Mathematical Logic 1 (1973), pp. 15–20.

[85] Pogorzelski W.A., Prucnal T., *Equivalence of the structural completeness theorem for propositional calculus and Boolean representation theorem*; Reports on Mathematical Logic 3 (1974), pp. 37–40.

[86] Pogorzelski W.A., Prucnal T., *Introduction to Mathematical Logic; Elements of the algebra of propositional logic*; Silesian University, Katowice (mimeographed) 1974.

[87] Pogorzelski W.A., Słupecki J., *Basic properties of deductive systems based on non–classical logics* (in Polish); Studia Logica 9 (1960), pp. 163–176.

[88] Pogorzelski W.A., Wojtylak P., *Elements of the Theory of Completeness in Propositional Logic*; Silesian University, Katowice 1982.

[89] Pogorzelski W.A., Wojtylak P., *Cn–definitions of propositional connectives*; Studia Logica 67 (2001), pp. 1–26.

[90] Pogorzelski W.A., Wojtylak P., *A proof system for classical logic*; Studia Logica 80 (2005), pp. 95–104.

[91] Połacik T., *Operators defined by propositional quantification and their interpretations over Cantor space*; Reports on Mathematical Logic 27 (1993), pp. 67–79.

[92] Połacik T., *Second order propositional operators over Cantor space*; Studia Logica 53 (1994), pp. 93–105.

[93] Połacik T., *Propositional quantification in the monadic fragment of intuitionistic logic*; Journal of Symbolic Logic 63 (1998), pp. 269–300.

[94] Połacik T., *Pitts' quantifiers are not propositional quantification*; Notre Dame Journal of Formal Logic 39 (1998), pp. 531-544.

[95] Połacik T., *Anti-chains, focuses and projective formulas*; Bulletin of the Section of Logic 34/1 (2005), pp. 1–12.

[96] Połacik T., *Partially-elementary extension Kripke models: a characterization and applications*; The Logic Journal of the IGPL 14/1 (2006), pp. 73-86.

[97] Post E., *Introduction to a general theory of elementary proposition*; American Journal of Mathematics 43 (1921), pp. 163–185.

[98] Prucnal T., *On the structural completeness of some pure implicational propositional calculi*; Studia Logica 30 (1972), pp. 45–52.

[99] Prucnal T., *Structural completeness of Lewis's systems S5*; Bulletin de l'Académie Polonaise des Sciences, Série des Sciences Mathématiques, Astronomiques et Physiques 29 (1972), pp. 101–103.

[100] Prucnal T, *Interpretations of the classical implicational calculus in nonclassical implicational calculi*; Studia Logica 33 (1974), pp. 59–63.

[101] Prucnal T., *Structural completeness of Medvedev's propositional calculus*; Reports on Mathematical Logic, 6 (1976), pp. 103–105.

[102] Prucnal T., *On two problems of Harvey Friedman*; Studia Logica, 38 no. 3 (1979), pp. 247–262.

[103] Prucnal T., *On the structural completeness of some pure implicational logics*; Salzburg 1983, in Foundation of Logic and Linguistic, Plenum New York 1985, pp. 31–41.

[104] Prucnal T., Wroński A. *An algebraic characterization of the notion of structural completeness*; Studia Logica, Bulletin of the Section of Logic 3 (1974) no. 3 (1979), pp. 30–33.

[105] Prucnal T., Wroński A., *An algebraic characterization of the notion of structural completeness*; Bulletin of the Section of Logic 3 (1977), pp. 30–33.

[106] Rasiowa H., *An algebraic approach to non–classical logics*; PWN–Polish Scientific Publishers Warszawa and North–Holland Publishing Company Amsterdam 1974.

[107] Rasiowa H., Sikorski R., *The mathematics of metamathematics*; PWN–Polish Scientific Publishers, Warszawa 1963 (3rd edition 1970).

[108] Rautenberg W., *2–element matrices*; Studia Logica 40 (1981), pp. 315–353.

[109] Rieger L., *On the lattice theory of Brouwerian propositional logic*; Acta Facultatis Rerum Naturalium Universitas Carolinae 189 (1949), pp. 1–40.

[110] Rose A., *The degree of completeness of the m–valued Łukasiewicz propositional calculus*; The Journal of the London Mathematical Society 27 (1952), pp. 92–102.

[111] Rybakov V.V., *Decidability of admissibility in the modal system Grz and intuitionistic logic*; Math. USSR Izvestya 50 (1986); English translation 28 (1987), pp. 589–608.

[112] Rybakov V.V., *Admissible Logical Inference Rules*; Studies in Logic and the Foundations of Mathematics, Vol.136, Elsevier Sci.Publ.,North-Holland, New-York and Amsterdam, 1997.

[113] Scroggs S.J., *Extensions of the Lewis system S5*; The Journal of Symbolic Logic 16 (1951), pp. 112–120.

[114] Shoesmith D.J., Smiley T.J., *Deducibility and many–valuedness*; The Journal of Symbolic Logic 36 (1971), pp. 610–622.

[115] Smiley T.J., *The independence of connectives*; The Journal of Symbolic Logic 26 (1962), pp. 426–436.

[116] Suszko R., *On filters and closure systems*; Bulletin of the Section of Logic 6 (1977), pp. 151–155.

[117] Tarski A., *Über einige fundamentale Begriffe der Metamathematik*; Comptes Rendus des Séances de la Societé des Sciences et des Lettres de Varsovie, Cl III, 23 (1930), pp. 22–29.

[118] Tarski A., *Fundamentale Begriffe der Methodologie der deductiven Wissenschaften* I; Monatshefte für Mathematik und Physik 37 (1930), pp. 361–404.

[119] Tarski A., *Über die Erweiterungen der unvollständigen systeme des Aussagenkalküls*; Ergebnisse eines mathematischen Kolloquiums 7 (1935), pp. 283–410.

[120] Thiele H., *Eine Axiomatisierung der zweiwertigen Prädikatenkalküle der ersten Stufe, welche die Implication entahlten*; Zeitschrift für Mathematische Logik und Grundlagen der Mathematik 2 (1956), pp. 93–106.

[121] Thomas I., *Finite limitations of Dummett's LC*; Notre Dame Journal of Formal Logic 3 (1962), pp. 170–174.

[122] Tokarz M., *On structural completeness of Łukasiewicz logics*; Studia Logica 30 (1972), pp. 53–58.

[123] Tokarz M., *Connections between some notions of completeness of structural propositional calculi*; Studia Logica 32 (1973), pp. 77–91.

[124] Tokarz M., *A strongly finite logic with infinite degree of maximality*; Studia Logica 35 (1976), pp. 447–451.

[125] Tokarz M., *The existence of matrices strongly adequate for E, R and their fragments*; Studia Logica 38 (1979), pp. 75–85.

[126] Tokarz M., *Maximality, Post–completeness and structural completeness*, 1979, (mimeographed).

[127] Troelstra A.S., *On intermediate propositional logic*; Indagationes Mathematicae 27 (1965), pp. 141–152.

[128] Tuziak R., *An axiomatization of the finitely–valued Łukasiewicz calculus*, Studia Logica 48 (1988), pp. 49–56.

[129] Wajsberg M., *Beiträge zum Metaaussagenkalkül I*; Monatshefte für Mathematik und Physik 42(1935), pp. 221–242; English translation *Contributions to meta–calculus of propositions I*, in Mordchaj Wajsberg, *Logical Works* ed. by S.J. Surma, Ossolineum, Warszawa–Wrocław 1973.

[130] Wojtylak P., *A new proof of structural completeness of Łukasiewicz's logics*; Bulletin of the Section of Logic 5 (1976), pp. 145–152.

[131] Wojtylak P., *On structural completeness of many–valued logics*; Studia Logica 37 (1978), pp. 139–147.

[132] Wojtylak P., *Strongly finite logics: finite axiomatizability and the problem of supremum*; Bulletin of the Section of Logic 8 (1979), pp. 99–111.

[133] Wojtylak P., *An example of a finite though finitely non–axiomatizable matrix*; Reports on Mathematical Logic 17 (1979), pp. 39–46.

[134] Wojtylak P., *Mutual interpretability of sentential logics II*; Reports on Mathematical Logic 12 (1981), pp. 51–66.

[135] Wojtylak P., *Corrections to the paper of T. Prucnal: Structural completeness of Lewis's system S5*; Reports on Mathematical Logic 15 (1983), pp. 67–70.

[136] Wojtylak P., *Independent axiomatizability of sets of sentences*; Annals of Pure and Applied Logic 44 (1989), pp. 259–299,

[137] Wojtylak P., *Structural completeness of implicational logics*; Studia Logica 51 (1991), pp. 275–298.

[138] Wojtylak P., *Axiomatizability of logical matrices* in *Emil L. Post and the Problem of Mechanical Provability, A survey of Posts's Contributions in the Century of His Birth*; Studies in Logic, Grammar and Rhetoric 2 (1998), pp. 77–89.

[139] Wojtylak P., *On a problem of H.Friedman and irs solution by T.Prucnal*; Reports on Mathematical Logic 38 (2004), pp. 69–86.

[140] Wójcicki R., *Some remarks on the consequence operation in sentential logics*; Fundamenta Mathematicae 68 (1970), pp. 269–279.

[141] Wójcicki R., *The degree of completeness of the finite–valued propositional calculus*; Prace z Logiki, Zeszyty Naukowe UJ 7 (1972), pp. 77–84.

[142] Wójcicki R., *Matrix approach in methodology of sentential calculi*; Studia Logica 32 (1973), pp. 7–39.

[143] Wójcicki R., *The logics stronger than Łukasiewicz's 3-valued calculus — the notion of degree of maximality versus the notion of degree of completeness*; Studia Logica 33 (1974), pp. 201–214.

[144] Wójcicki R., *On reconstructability of the classical propositional logic in intuitionistic logic*; Bulletin de l'Académie Polonaise des Sciences, série des Sciences Mathématiques, Astronomiques et Physiques 9 (1970), pp. 421–422.

[145] Wójcicki R., *On matrix representations of consequence operations of Łukasiewicz sentential calculi*; Zeitshrift für Mathematische Logik und Grundlagen der Mathematik 19 (1973), pp. 230–247.

[146] Wójcicki R., *Strongly finite sentential calculi*; in *Selected Papers on Łukasiewicz Sentential Calculi* ed. R. Wójcicki and G. Malinowski, Ossolineum, Warszawa–Wrocław 1977.

[147] Wójcicki R., *Theory of logical calculi*; Synthese Library vol. 199, Kluwer Academic Publishers 1988.

[148] Wroński A., *Intermediate logics and the disjunction property*; Reports on Mathematical Logic 1 (1973), pp. 39–52.

[149] Wroński A., *The degree of completeness of some fragments of the intuitionistic propositional logic*; Reports on Mathematical Logic 2 (1974), pp. 55–62.

[150] Wroński A., *Remarks on intermediate logics with axioms containing only one variable*; Reports on Mathematical Logic 2 (1974), pp. 63–76.

[151] Wroński A., *On cardinalities of matrices strongly adequate for the intuitionistic propositional logic*; Reports on Mathematical Logic 3 (1974), pp. 67–72.

[152] Wroński A., *On finitely based consequence operations*. Studia Logica 35 (1976), pp. 453–458.

[153] Wroński A., *A three element matrix whose consequence is not finitely axiomatizable*; Studia Logica 35 (1979), pp. 453–458.

[154] Zaionc M., *On the asymptotic density of tautologies in logic of implication and negation*; Reports on Mathematical Logic 39 (2005), pp. 67–87.

[155] Zygmunt J., *An essay in matrix semantics for consequence relation*; Acta Universitatis Wratislaviensis No 741, Wrocław 1984.

[156] Zygmunt J., *A note on direct products and ultraproducts of logical matrices*; Studia Logica, 33 (1974), pp. 251–259.

Notation

Index

STUDIES IN UNIVERSAL LOGIC

Institution-independent Model Theory

Diaconescu, R., Institute of Mathematics „Simion Stoilow" of the Romanian Academy, Romania

BIRKHÄUSER

A model theory that is independent of any concrete logical system allows a general handling of a large variety of logics. This generality can be achieved by applying the theory of institutions that provides a precise mathematical formulation for the intuitive concept of a logical system. Especially in computer science, where the development of a huge number of specification logics is observable, institution-independent model theory simplifies and sometimes even enables a concise model-theoretic analysis of the system. Besides incorporating important methods and concepts from conventional model theory, the proposed top-down methodology allows for a structurally clean understanding of model-theoretic phenomena. As a consequence, results from conventional concrete model theory can be understood more easily, and sometimes even new results are obtained.

2008. Approx. 360 p. Softcover
ISBN 978-3-7643-8707-5
SUL — Studies in Universal Logic

www.birkhauser.ch

JOURNAL

Logica Universalis

First published in 2007
1 volume per year, 2 issues per volume, approx.
300 pages per volume
Format: 17 x 24 cm
ISSN 1661-8297 (print)
ISSN 1661-8300 (electronic)

BIRKHÄUSER

Editor-in-Chief
Jean-Yves Béziau
Swiss National Science Foundation
Switzerland
jean-yves.beziau@unine.ch

Logica Universalis (LU) publishes peer-reviewed
research papers related to universal features of
logics. Topics include general tools and techniques for
studying already existing logics and building new ones,
the study of classes of logics, the scope of validity and
the domain of application of fundamental theorems,
and also philosophical and historical aspects of
general concepts of logic.

Abstracted/Indexed in:
MathSciNet, SCOPUS, Zentralblatt Math

www.birkhauser.ch/LU